Aerospace Engineering

Aerospace Engineering

Edited by
Killian Sullivan

Larsen & Keller
www.larsen-keller.com

Aerospace Engineering
Edited by Killian Sullivan
ISBN: 978-1-63549-015-2 (Hardback)

© 2017 Larsen & Keller

 Larsen & Keller

Published by Larsen and Keller Education,
5 Penn Plaza,
19th Floor,
New York, NY 10001, USA

Cataloging-in-Publication Data

Aerospace engineering / edited by Killian Sullivan.
     p. cm.
Includes bibliographical references and index.
ISBN 978-1-63549-015-2
1. Aerospace engineering. 2. Aeronautics. I. Sullivan, Killian.
TL545 .A37 2017
629.1--dc23

The publisher's policy is to use permanent paper from mills that operate a sustainable forestry policy. Furthermore, the publisher ensures that the text paper and cover boards used have met acceptable environmental accreditation standards.

Printed and bound in the United States of America.

For more information regarding Larsen and Keller Education and its products, please visit the publisher's website www.larsen-keller.com

Table of Contents

Permissions

Index

Preface

Aerospace engineering is a vast field of engineering which is concerned with the design, construction and development of spacecrafts and aircrafts. It is divided into two major subfields namely, astronautical engineering and aeronautical engineering. This field deals with the economic as well as engineering aspect of aerospace and aircraft engineering. The book on aerospace engineering elucidates the concepts and innovative models around prospective developments with respect to aerospace engineering. It picks up individual branches and explains their need and contribution in the context of growth of this subject. The various sub-fields of aerospace engineering along with their technological progress that have future implications are glanced at in this text. As this field is emerging at a rapid pace, the contents of this textbook will help the students understand the modern concepts and applications of the subject.

To facilitate a deeper understanding of the contents of this book a short introduction of every chapter is written below:

Chapter 1- Aerospace is a branch of engineering which concerns itself with the art and science of flying in the Earth's atmosphere as well as outer space. It is a very diverse subject, with a number of industrial and military applications. This chapter provides an integrated understanding on aerospace engineering.

Chapter 2- Aerodynamics is the study of the motion of air, mainly when the interaction occurs with solid objects. Aerodynamics is similar to gas dynamics but the only difference being that gas dynamics applies to the study of all gases and is not only limited to air. Aerodynamics, control engineering, aeroelasticity, reaction engine and flight dynamics of spacecrafts and aircrafts are some of the key elements of aerospace engineering that have been discussed in the following text.

Chapter 3- The electronic systems that are used on aircrafts and spacecrafts are known as avionics. Devices that are used for controlling aircrafts, missiles and satellites are known as guidance systems. A guidance system also helps in the navigation of these objects. The aspects elucidated in this section are of vital importance, and it provides a better understanding of aerospace engineering.

Chapter 4- The processes involved in the designing of aircrafts are termed as aircraft manufacturing process. It usually depends on factors such as customer demand, safety protocols and economic constraints. To provide a better understanding on the topic, a brief explanation on gull wing, aircraft flight control system and aeronautics is also given. This chapter is a compilation of the various branches of aircraft manufacturing that form an integral part of the broader subject matter.

Chapter 5- Data is gathered during the flight of an aircraft. The collection of the data and the analysis of it is known as flight test. Cooper-Harper rating scale, flying qualities, drop test and airframe are some of the topics discussed in the following content. The aspects elucidated in this chapter are of vital importance, and provide a better understanding of aerospace engineering.

Chapter 6- Aircrafts are machines which are able to fly by the support of air. A monoplane has a single main wing place, in difference to a biplane. Zeppelin, monoplane, rotorcraft, helicopter and powered aircraft are some of the aircrafts which have been explained in the chapter. Aircrafts and helicopters are one of the significant and important topics related to aerospace engineering. The following chapter unfolds its crucial aspects in a critical yet systematic manner.

I owe the completion of this book to the never-ending support of my family, who supported me throughout the project.

Editor

Introduction to Aerospace Engineering

Aerospace is a branch of engineering which concerns itself with the art and science of flying in the Earth's atmosphere as well as outer space. It is a very diverse subject, with a number of industrial and military applications. This chapter provides an integrated understanding on aerospace engineering.

Aerospace Engineering

Aerospace engineering is the primary field of engineering concerned with the development of aircraft and spacecraft. It is divided into two major and overlapping branches: aeronautical engineering and astronautical engineering.

Aeronautical engineering was the original term for the field. As flight technology advanced to include craft operating in outer space, the broader term "aerospace engineering" has largely replaced it in common usage. Aerospace engineering, particularly the astronautics branch, is often colloquially referred to as "rocket science".

Overview

Flight vehicles are subjected to demanding conditions such as those produced by changes in atmospheric pressure and temperature, with structural loads applied upon vehicle components. Consequently, they are usually the products of various technological and engineering disciplines including aerodynamics, propulsion, avionics, materials science, structural analysis and manufacturing. The interaction between these technologies is known as aerospace engineering. Because of the complexity and number of disciplines involved, aerospace engineering is carried out by teams of engineers, each having their own specialized area of expertise.

History

The origin of aerospace engineering can be traced back to the aviation pioneers around the late 19th to early 20th centuries, although the work of Sir George Cayley dates from the last decade of the 18th to mid-19th century. One of the most important people in the history of aeronautics, Cayley was a pioneer in aeronautical engineering and is credited as the first person to separate the forces of lift and drag, which are in effect on any flight vehicle. Early knowledge of aeronautical engineering was largely empirical with some concepts and skills imported from other branches of engineering. Scientists understood some key elements of aerospace engineering, like fluid dynamics, in the 18th century. Many years later after the successful flights by the Wright brothers, the 1910s saw the development of aeronautical engineering through the design of World War I military aircraft.

Orville and Wilbur Wright flew the Wright Flyer in 1903 at Kitty Hawk, North Carolina.

The first definition of aerospace engineering appeared in February 1958. The definition considered the Earth's atmosphere and the outer space as a single realm, thereby encompassing both aircraft (*aero*) and spacecraft (*space*) under a newly coined word *aerospace*. In response to the USSR launching the first satellite, Sputnik into space on October 4, 1957, U.S. aerospace engineers launched the first American satellite on January 31, 1958. The National Aeronautics and Space Administration was founded in 1958 as a response to the Cold War.

Elements

Wernher von Braun, with the F-1 engines of the Saturn V first stage at the US Space and Rocket Center

Soyuz TMA-14M spacecraft engineered for descent by parachute

A fighter jet engine undergoing testing. The tunnel behind the engine allows noise and exhaust to escape.

Some of the elements of aerospace engineering are:

- Radar cross-section – the study of vehicle signature apparent to Radar remote sensing.

- Fluid mechanics – the study of fluid flow around objects. Specifically aerodynamics concerning the flow of air over bodies such as wings or through objects such as wind tunnels.

- Astrodynamics – the study of orbital mechanics including prediction of orbital elements when given a select few variables. While few schools in the United States teach this at the undergraduate level, several have graduate programs covering this topic (usually in conjunction with the Physics department of said college or university).

- Statics and Dynamics (engineering mechanics) – the study of movement, forces, moments in mechanical systems.

- Mathematics – in particular, calculus, differential equations, and linear algebra.

- Electrotechnology – the study of electronics within engineering.

- Propulsion – the energy to move a vehicle through the air (or in outer space) is provided by internal combustion engines, jet engines and turbomachinery, or rockets. A more recent addition to this module is electric propulsion and ion propulsion.

- Control engineering – the study of mathematical modeling of the dynamic behavior of systems and designing them, usually using feedback signals, so that their dynamic behavior is desirable (stable, without large excursions, with minimum error). This applies to the dynamic behavior of aircraft, spacecraft, propulsion systems, and subsystems that exist on aerospace vehicles.

- Aircraft structures – design of the physical configuration of the craft to withstand the forces encountered during flight. Aerospace engineering aims to keep structures lightweight and low-cost, while maintaining structural integrity.

- Materials science – related to structures, aerospace engineering also studies the materials of which the aerospace structures are to be built. New materials with very specific properties are invented, or existing ones are modified to improve their performance.

- Solid mechanics – Closely related to material science is solid mechanics which deals with stress and strain analysis of the components of the vehicle. Nowadays there are several Finite Element programs such as MSC Patran/Nastran which aid engineers in the analytical process.

- Aeroelasticity – the interaction of aerodynamic forces and structural flexibility, potentially causing flutter, divergence, etc.

- Avionics – the design and programming of computer systems on board an aircraft or spacecraft and the simulation of systems.

- Software – the specification, design, development, test, and implementation of computer software for aerospace applications, including flight software, ground control software, test & evaluation software, etc.

- Risk and reliability – the study of risk and reliability assessment techniques and the mathematics involved in the quantitative methods.

- Noise control – the study of the mechanics of sound transfer.

- Aeroacoustics – the study of noise generation via either turbulent fluid motion or aerodynamic forces interacting with surfaces.

- Flight test – designing and executing flight test programs in order to gather and analyze performance and handling qualities data in order to determine if an aircraft meets its design and performance goals and certification requirements.

The basis of most of these elements lies in theoretical physics, such as fluid dynamics for aerodynamics or the equations of motion for flight dynamics. There is also a large empirical component. Historically, this empirical component was derived from testing of scale models and prototypes, either in wind tunnels or in the free atmosphere. More recently, advances in computing have enabled the use of computational fluid dynamics to simulate the behavior of fluid, reducing time and expense spent on wind-tunnel testing. Those studying hydrodynamics or Hydroacoustics often obtained degrees in Aerospace Engineering.

Additionally, aerospace engineering addresses the integration of all components that constitute an aerospace vehicle (subsystems including power, aerospace bearings, communications, thermal control, life support, etc.) and its life cycle (design, temperature, pressure, radiation, velocity, lifetime).

Degree Programs

Aerospace engineering may be studied at the advanced diploma, bachelor's, master's, and Ph.D. levels in aerospace engineering departments at many universities, and in mechanical engineering departments at others. A few departments offer degrees in space-focused astronautical engineering. Some institutions differentiate between aeronautical and astronautical engineering. Graduate degrees are offered in advanced or specialty areas for the aerospace industry.

A background in chemistry, physics, computer science and mathematics is important for students pursuing an aerospace engineering degree.

In Popular Culture

The term "rocket scientist" is sometimes used to describe a person of great intelligence since "rocket science" is seen as a practice requiring great mental ability, especially technical and mathematical ability. The term is used ironically in the expression "It's not rocket science" to indicate that a task is simple. Strictly speaking, the use of "science" in "rocket science" is a misnomer since science is about understanding the origins, nature, and behavior of the universe; engineering is about using scientific and engineering principles to solve problems and develop new technology. However, the media and the public often use "science" and "engineering" as synonyms.

Key Elements of Aerospace Engineering

Aerodynamics is the study of the motion of air, mainly when the interaction occurs with solid objects. Aerodynamics is similar to gas dynamics but the only difference being that gas dynamics applies to the study of all gases and is not only limited to air. Aerodynamics, control engineering, aeroelasticity, reaction engine and flight dynamics of spacecrafts and aircrafts are some of the key elements of aerospace engineering that have been discussed in the following text.

Aerodynamics

A vortex is created by the passage of an aircraft wing, revealed by smoke. Vortices are one of the many phenomena associated with the study of aerodynamics.

Aerodynamics, from Greek ἀήρ *aer* (air) + δυναμική (dynamics), is a branch of fluid dynamics concerned with studying the motion of air, particularly when it interacts with a solid object, such as an airplane wing. Aerodynamics is a sub-field of fluid dynamics and gas dynamics, and many aspects of aerodynamics theory are common to these fields. The term *aerodynamics* is often used synonymously with gas dynamics, with the difference being that "gas dynamics" applies to the study of the motion of all gases, not limited to air. Formal aerodynamics study in the modern sense began in the eighteenth century, although observations of fundamental concepts such as aerodynamic drag have been recorded much earlier. Most of the early efforts in aerodynamics worked towards achieving heavier-than-air flight, which was first demonstrated by Wilbur and

Orville Wright in 1903. Since then, the use of aerodynamics through mathematical analysis, empirical approximations, wind tunnel experimentation, and computer simulations has formed the scientific basis for ongoing developments in heavier-than-air flight and a number of other technologies. Recent work in aerodynamics has focused on issues related to compressible flow, turbulence, and boundary layers and has become increasingly computational in nature.

History

Modern aerodynamics only dates back to the seventeenth century, but aerodynamic forces have been harnessed by humans for thousands of years in sailboats and windmills, and images and stories of flight appear throughout recorded history, such as the Ancient Greek legend of Icarus and Daedalus. Fundamental concepts of continuum, drag, and pressure gradients appear in the work of Aristotle and Archimedes.

In 1726, Sir Isaac Newton became the first person to develop a theory of air resistance, making him one of the first aerodynamicists. Dutch-Swiss mathematician Daniel Bernoulli followed in 1738 with *Hydrodynamica* in which he described a fundamental relationship between pressure, density, and flow velocity for incompressible flow known today as Bernoulli's principle, which provides one method for calculating aerodynamic lift. In 1757, Leonhard Euler published the more general Euler equations which could be applied to both compressible and incompressible flows. The Euler equations were extended to incorporate the effects of viscosity in the first half of the 1800s, resulting in the Navier-Stokes equations. The Navier-Stokes equations are the most general governing equations of fluid flow and are difficult to solve.

A replica of the Wright brothers' wind tunnel is on display at the Virginia Air and Space Center. Wind tunnels were key in the development and validation of the laws of aerodynamics.

In 1799, Sir George Cayley became the first person to identify the four aerodynamic forces of flight (weight, lift, drag, and thrust), as well as the relationships between them, outlining the work towards achieving heavier-than-air flight for the next century. In 1871, Francis Herbert Wenham constructed the first wind tunnel, allowing precise measurements of aerodynamic forces. Drag theories were developed by Jean le Rond d'Alembert, Gustav Kirchhoff, and Lord Rayleigh. In 1889, Charles Renard, a French aeronautical engineer, became the first person to reasonably predict the power needed for sustained flight. Otto Lilienthal, the first person to become highly successful with glider flights, was also the first to propose thin, curved airfoils that would produce high lift and low drag. Building on these developments as well as research carried out in their own wind tunnel, the Wright brothers flew the first powered airplane on December 17, 1903.

During the time of the first flights, Frederick W. Lanchester, Martin Wilhelm Kutta, and Nikolai Zhukovsky independently created theories that connected circulation of a fluid flow to lift. Kutta and Zhukovsky went on to develop a two-dimensional wing theory. Expanding upon the work of Lanchester, Ludwig Prandtl is credited with developing the mathematics behind thin-airfoil and lifting-line theories as well as work with boundary layers.

As aircraft speed increased, designers began to encounter challenges associated with air compressibility at speeds near or greater than the speed of sound. The differences in air flows under these conditions led to problems in aircraft control, increased drag due to shock waves, and structural dangers due to aeroelastic flutter. The ratio of the flow speed to the speed of sound was named the Mach number after Ernst Mach who was one of the first to investigate the properties of supersonic flow. William John Macquorn Rankine and Pierre Henri Hugoniot independently developed the theory for flow properties before and after a shock wave, while Jakob Ackeret led the initial work on calculating the lift and drag of supersonic airfoils. Theodore von Kármán and Hugh Latimer Dryden introduced the term transonic to describe flow speeds around Mach 1 where drag increases rapidly. This rapid increase in drag led aerodynamicists and aviators to disagree on whether supersonic flight was achievable until the sound barrier was broken for the first time in 1947 using the Bell X-1 aircraft.

By the time the sound barrier was broken, much of the subsonic and low supersonic aerodynamics knowledge had matured. The Cold War fueled an ever evolving line of high performance aircraft. Computational fluid dynamics began as an effort to solve for flow properties around complex objects and has rapidly grown to the point where entire aircraft can be designed using a computer, with wind-tunnel tests followed by flight tests to confirm the computer predictions. Knowledge of supersonic and hypersonic aerodynamics has also matured since the 1960s, and the goals of aerodynamicists have shifted from understanding the behavior of fluid flow to understanding how to engineer a vehicle to interact appropriately with the fluid flow. Designing aircraft for supersonic and hypersonic conditions, as well as the desire to improve the aerodynamic efficiency of current aircraft and propulsion systems, continues to fuel new research in aerodynamics, while work continues to be done on important problems in basic aerodynamic theory related to flow turbulence and the existence and uniqueness of analytical solutions to the Navier-Stokes equations.

Fundamental Concepts

Understanding the motion of air around an object (often called a flow field) enables the calculation of forces and moments acting on the object. In many aerodynamics problems, the forces of interest are the fundamental forces of flight: lift, drag, thrust, and weight. Of these, lift and drag are aerodynamic forces, i.e. forces due to air flow over a solid body. Calculation of these quantities is often founded upon the assumption that the flow field behaves as a continuum. Continuum flow fields are characterized by properties such as flow velocity, pressure, density, and temperature, which may be functions of spatial position and time. These properties may be directly or indirectly measured in aerodynamics experiments or calculated from equations for the conservation of mass, momentum, and energy in air flows. Density, flow velocity, and an additional property, viscosity, are used to classify flow fields.

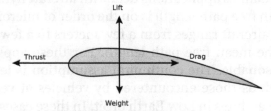

Forces of flight on an airfoil

Flow Classification

Flow velocity is used to classify flows according to speed regime. Subsonic flows are flow fields in which the air speed field is always below the local speed of sound. Transonic flows include both regions of subsonic flow and regions in which the local flow speed is greater than the local speed of sound. Supersonic flows are defined to be flows in which the flow speed is greater than the speed of sound everywhere. A fourth classification, hypersonic flow, refers to flows where the flow speed is much greater than the speed of sound. Aerodynamicists disagree on the precise definition of hypersonic flow.

Compressibility refers to whether or not the flow in a problem can have a varying density. Subsonic flows are often assumed to be incompressible, i.e. the density is assumed to be constant. Transonic and supersonic flows are compressible, and neglecting the changes in density in these flow fields will yield inaccurate results when performing calculations.

Viscosity is associated with the frictional forces in a flow. In some flow fields, viscous effects are very small, and approximate solutions may safely neglect viscous effects. These approximations are called inviscid flows. Flows for which viscosity is not neglected are called viscous flows. Finally, aerodynamic problems may also be classified by the flow environment. External aerodynamics is the study of flow around solid objects of various shapes (e.g. around an airplane wing), while internal aerodynamics is the study of flow through passages inside solid objects (e.g. through a jet engine).

Continuum Assumption

Unlike liquids and solids, gases are composed of discrete molecules which occupy only a small fraction of the volume filled by the gas. On a molecular level, flow fields are made up of many individual collisions between gas molecules and between gas molecules and solid surfaces. However, in most aerodynamics applications, the discrete molecular nature of gases is ignored, and the flow field is assumed to behave as a continuum. This assumption allows fluid properties such as density and flow velocity to be defined everywhere within the flow.

Validity of the continuum assumption is dependent on the density of the gas and the application in question. For the continuum assumption to be valid, the mean free path length must be much smaller than the length scale of the application in question. For example, many aerodynamics applications deal with aircraft flying in atmospheric conditions, where the mean free path length is on the order of micrometers. In these cases, the length scale of the aircraft ranges from a few meters to a few tens of meters, which is much larger than the mean free path length. For these applications, the continuum assumption is reasonable. The continuum assumption is less valid for extremely low-density flows, such as those encountered by vehicles at very high altitudes (e.g. 300,000 ft/90 km) or satellites in Low Earth orbit. In these cases, statistical mechanics is a more accurate method of solving the problem than continuous aerodynamics. The Knudsen number can be used to guide the choice between statistical mechanics and the continuous formulation of aerodynamics.

Conservation Laws

Aerodynamic problems are typically solved using fluid dynamics conservation laws as applied to a fluid continuum. Three conservation principles are used:

1. Conservation of mass: In fluid dynamics, the mathematical formulation of this principle is known as the mass continuity equation, which requires that mass is neither created nor destroyed within a flow of interest.

2. Conservation of momentum: In fluid dynamics, the mathematical formulation of this principle can be considered an application of Newton's Second Law. Momentum within a flow of interest is only created or destroyed due to the work of external forces, which may include both surface forces, such as viscous (frictional) forces, and body forces, such as weight. The momentum conservation principle may be expressed as either a single vector equation or a set of three scalar equations, derived from the components of the three-dimensional flow velocity vector. In its most complete form, the momentum conservation equations are known as the Navier-Stokes equations. The Navier-Stokes equations have no known analytical solution and are solved in modern aerodynamics using computational techniques. Because of the computational cost of solving

these complex equations, simplified expressions of momentum conservation may be appropriate for specific applications. The Euler equations are a set of momentum conservation equations which neglect viscous forces used widely by modern aerodynamicists in cases where the effect of viscous forces is expected to be small. Additionally, Bernoulli's equation is a solution to the momentum conservation equation of an inviscid flow, neglecting gravity.

3. Conservation of energy: The energy conservation equation states that energy is neither created nor destroyed within a flow, and that any addition or subtraction of energy is due either to the fluid flow in and out of the region of interest, heat transfer, or work.

The ideal gas law or another equation of state is often used in conjunction with these equations to form a determined system to solve for unknown variables.

Branches of Aerodynamics

Aerodynamic problems are classified by the flow environment or properties of the flow, including flow speed, compressibility, and viscosity. *External* aerodynamics is the study of flow around solid objects of various shapes. Evaluating the lift and drag on an airplane or the shock waves that form in front of the nose of a rocket are examples of external aerodynamics. *Internal* aerodynamics is the study of flow through passages in solid objects. For instance, internal aerodynamics encompasses the study of the airflow through a jet engine or through an air conditioning pipe.

Aerodynamic problems can also be classified according to whether the flow speed is below, near or above the speed of sound. A problem is called subsonic if all the speeds in the problem are less than the speed of sound, transonic if speeds both below and above the speed of sound are present (normally when the characteristic speed is approximately the speed of sound), supersonic when the characteristic flow speed is greater than the speed of sound, and hypersonic when the flow speed is much greater than the speed of sound. Aerodynamicists disagree over the precise definition of hypersonic flow; a rough definition considers flows with Mach numbers above 5 to be hypersonic.

The influence of viscosity in the flow dictates a third classification. Some problems may encounter only very small viscous effects on the solution, in which case viscosity can be considered to be negligible. The approximations to these problems are called inviscid flows. Flows for which viscosity cannot be neglected are called viscous flows.

Incompressible Aerodynamics

An incompressible flow is a flow in which density is constant in both time and space. Although all real fluids are compressible, a flow problem is often considered incompressible if the effect of the density changes in the problem on the outputs of interest is small. This is more likely to be true when the flow speeds are significantly lower than the speed of sound. Effects of compressibility are more significant at speeds close to or above the speed of sound. The Mach number is used to evaluate whether the incompressibility can be assumed or the flow must be solved as compressible.

Subsonic Flow

Subsonic (or low-speed) aerodynamics studies fluid motion in flows which are much lower than the speed of sound everywhere in the flow. There are several branches of subsonic flow but one special case arises when the flow is inviscid, incompressible and irrotational. This case is called potential flow and allows the differential equations used to be a simplified version of the governing equations of fluid dynamics, thus making available to the aerodynamicist a range of quick and easy solutions.

In solving a subsonic problem, one decision to be made by the aerodynamicist is whether to incorporate the effects of compressibility. Compressibility is a description of the amount of change of density in the problem. When the effects of compressibility on the solution are small, the aerodynamicist may choose to assume that density is constant. The problem is then an incompressible low-speed aerodynamics problem. When the density is allowed to vary, the problem is called a compressible problem. In air, compressibility effects are usually ignored when the Mach number in the flow does not exceed 0.3 (about 335 feet (102 m) per second or 228 miles (366 km) per hour at 60 °F (16 °C)). Above 0.3, the problem should be solved by using compressible aerodynamics.

Compressible Aerodynamics

According to the theory of aerodynamics, a flow is considered to be compressible if its change in density with respect to pressure is non-zero along a streamline. This means that – unlike incompressible flow – changes in density must be considered. In general, this is the case where the Mach number in part or all of the flow exceeds 0.3. The Mach .3 value is rather arbitrary, but it is used because gas flows with a Mach number below that value demonstrate changes in density with respect to the change in pressure of less than 5%. Furthermore, that maximum 5% density change occurs at the stagnation point of an object immersed in the gas flow and the density changes around the rest of the object will be significantly lower. Transonic, supersonic, and hypersonic flows are all compressible.

Transonic Flow

The term Transonic refers to a range of flow velocities just below and above the local speed of sound (generally taken as Mach 0.8–1.2). It is defined as the range of speeds between the critical Mach number, when some parts of the airflow over an aircraft become supersonic, and a higher speed, typically near Mach 1.2, when all of the airflow is supersonic. Between these speeds, some of the airflow is supersonic, while some of the airflow is not supersonic.

Supersonic Flow

Supersonic aerodynamic problems are those involving flow speeds greater than the speed of sound. Calculating the lift on the Concorde during cruise can be an example of a supersonic aerodynamic problem.

Supersonic flow behaves very differently from subsonic flow. Fluids react to differences in pressure; pressure changes are how a fluid is "told" to respond to its environment. Therefore, since sound is in fact an infinitesimal pressure difference propagating through a fluid, the speed of sound in that fluid can be considered the fastest speed that "information" can travel in the flow. This difference most obviously manifests itself in the case of a fluid striking an object. In front of that object, the fluid builds up a stagnation pressure as impact with the object brings the moving fluid to rest. In fluid traveling at subsonic speed, this pressure disturbance can propagate upstream, changing the flow pattern ahead of the object and giving the impression that the fluid "knows" the object is there and is avoiding it. However, in a supersonic flow, the pressure disturbance cannot propagate upstream. Thus, when the fluid finally does strike the object, it is forced to change its properties – temperature, density, pressure, and Mach number—in an extremely violent and irreversible fashion called a shock wave. The presence of shock waves, along with the compressibility effects of high-flow velocity fluids, is the central difference between supersonic and subsonic aerodynamics problems.

Hypersonic Flow

In aerodynamics, hypersonic speeds are speeds that are highly supersonic. In the 1970s, the term generally came to refer to speeds of Mach 5 (5 times the speed of sound) and above. The hypersonic regime is a subset of the supersonic regime. Hypersonic flow is characterized by high temperature flow behind a shock wave, viscous interaction, and chemical dissociation of gas.

Associated Terminology

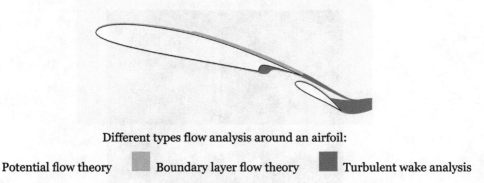

Different types flow analysis around an airfoil:

Potential flow theory   Boundary layer flow theory   Turbulent wake analysis

The incompressible and compressible flow regimes produce many associated phenomena, such as boundary layers and turbulence.

Boundary Layers

The concept of a boundary layer is important in many aerodynamic problems. The viscosity and fluid friction in the air is approximated as being significant only in this thin layer. This principle makes aerodynamics much more tractable mathematically.

Turbulence

In aerodynamics, turbulence is characterized by chaotic, stochastic property changes in the flow. This includes low momentum diffusion, high momentum convection, and rapid variation of pressure and flow velocity in space and time. Flow that is not turbulent is called laminar flow.

Aerodynamics in other Fields

Aerodynamics is important in a number of applications other than aerospace engineering. It is a significant factor in any type of vehicle design, including automobiles. It is important in the prediction of forces and moments in sailing. It is used in the design of mechanical components such as hard drive heads. Structural engineers also use aerodynamics, and particularly aeroelasticity, to calculate wind loads in the design of large buildings and bridges. Urban aerodynamics seeks to help town planners and designers improve comfort in outdoor spaces, create urban microclimates and reduce the effects of urban pollution. The field of environmental aerodynamics studies the ways atmospheric circulation and flight mechanics affect ecosystems. The aerodynamics of internal passages is important in heating/ventilation, gas piping, and in automotive engines where detailed flow patterns strongly affect the performance of the engine. People who do wind turbine design use aerodynamics. A few aerodynamic equations are used as part of numerical weather prediction.

Control Engineering

Control systems play a critical role in space flight

Control engineering or control systems engineering is the engineering discipline that applies control theory to design systems with desired behaviors. The practice uses sensors to measure the output performance of the device being controlled and those measurements can be used to give feedback to the input actuators that can make corrections toward desired performance. When a device is designed to perform without the need of human inputs for correction it is called automatic control (such as cruise control for regulating the speed of a car). Multi-disciplinary in nature, control systems engineering activities focus on implementation of control systems mainly derived by mathematical modeling of systems of a diverse range.

Overview

Modern day control engineering is a relatively new field of study that gained significant attention during the 20th century with the advancement of technology. It can be broadly defined or classified as practical application of control theory. Control engineering has an essential role in a wide range of control systems, from simple household washing machines to high-performance F-16 fighter aircraft. It seeks to understand physical systems, using mathematical modeling, in terms of inputs, outputs and various components with different behaviors, use control systems design tools to develop controllers for those systems and implement controllers in physical systems employing available technology. A system can be mechanical, electrical, fluid, chemical, financial and even biological, and the mathematical modeling, analysis and controller design uses control theory in one or many of the time, frequency and complex-s domains, depending on the nature of the design problem.

History

Automatic control systems were first developed over two thousand years ago. The first feedback control device on record is thought to be the ancient Ktesibios's water clock in Alexandria, Egypt around the third century B.C. It kept time by regulating the water level in a vessel and, therefore, the water flow from that vessel. This certainly was a successful device as water clocks of similar design were still being made in Baghdad when the Mongols captured the city in 1258 A.D. A variety of automatic devices have been used over the centuries to accomplish useful tasks or simply to just entertain. The latter includes the automata, popular in Europe in the 17th and 18th centuries, featuring dancing figures that would repeat the same task over and over again; these automata are examples of open-loop control. Milestones among feedback, or "closed-loop" automatic control devices, include the temperature regulator of a furnace attributed to Drebbel, circa 1620, and the centrifugal flyball governor used for regulating the speed of steam engines by James Watt in 1788.

In his 1868 paper "On Governors", James Clerk Maxwell was able to explain instabilities exhibited by the flyball governor using differential equations to describe the control system. This demonstrated the importance and usefulness of mathematical models and methods in understanding complex phenomena, and signaled the beginning of mathematical control and systems theory. Elements of control theory had appeared earlier but not as dramatically and convincingly as in Maxwell's analysis.

Control theory made significant strides over the next century. New mathematical techniques as well as advancements in electronic and computer technologies made it possible to control significantly more complex dynamical systems than the original flyball governor could stabilize. New mathematical techniques included developments in optimal control in the 1950s and 1960s followed by progress in stochastic, robust, adaptive, nonlinear, and azid-based control methods in the 1970s and 1980s. Applications of control methodology have helped to make possible space travel and communication satellites, safer and more efficient aircraft, cleaner automobile engines, and cleaner and more-efficient chemical processes.

Before it emerged as a unique discipline, control engineering was practiced as a part of mechanical engineering and control theory was studied as a part of electrical engineering since electrical circuits can often be easily described using control theory techniques. In the very first control relationships, a current output was represented with a voltage control input. However, not having proper technology to implement electrical control systems, designers left with the option of less efficient and slow responding mechanical systems. A very effective mechanical controller that is still widely used in some hydro plants is the governor. Later on, previous to modern power electronics, process control systems for industrial applications were devised by mechanical engineers using pneumatic and hydraulic control devices, many of which are still in use today.

Control Theory

There are two major divisions in control theory, namely, classical and modern, which have direct implications over the control engineering applications. The scope of classical control theory is limited to single-input and single-output (SISO) system design, except when analyzing for disturbance rejection using a second input. The system analysis is carried out in the time domain using differential equations, in the complex-s domain with the Laplace transform, or in the frequency domain by transforming from the complex-s domain. Many systems may be assumed to have a second order and single variable system response in the time domain. A controller designed using classical theory often requires on-site tuning due to incorrect design approximations. Yet, due to the easier physical implementation of classical controller designs as compared to systems designed using modern control theory, these controllers are preferred in most industrial applications. The most common controllers designed using classical control theory are PID controllers. A less common implementation may include either or both a Lead or Lag filter. The ultimate end goal is to meet a requirements set typically provided in the time-domain called the Step response, or at times in the frequency domain called the Open-Loop response. The Step response characteristics applied in a specification are typically percent overshoot, settling time, etc. The Open-Loop response characteristics applied in a specification are typically Gain and Phase margin and bandwidth. These characteristics may be evaluated through simulation including a dynamic model of the system under control coupled with the compensation model.

In contrast, modern control theory is carried out in the state space, and can deal with multiple-input and multiple-output (MIMO) systems. This overcomes the limitations of classical control theory in more sophisticated design problems, such as fighter aircraft control, with the limitation that no frequency domain analysis is possible. In modern design, a system is represented to the greatest advantage as a set of decoupled first order differential equations defined using state variables. Nonlinear, multivariable, adaptive and robust control theories come under this division. Matrix methods are significantly limited for MIMO systems where linear independence cannot be assured in the relationship between inputs and outputs. Being fairly new, modern control theory has many areas yet to be explored. Scholars like Rudolf E. Kalman and Aleksandr Lyapunov are well-known among the people who have shaped modern control theory.

Control Systems

Control engineering is the engineering discipline that focuses on the modeling of a diverse range of dynamic systems (e.g. mechanical systems) and the design of controllers that will cause these systems to behave in the desired manner. Although such controllers need not be electrical many are and hence control engineering is often viewed as a subfield of electrical engineering. However, the falling price of microprocessors is making the actual implementation of a control system essentially trivial. As a result, focus is shifting back to the mechanical and process engineering discipline, as intimate knowledge of the physical system being controlled is often desired.

Electrical circuits, digital signal processors and microcontrollers can all be used to implement control systems. Control engineering has a wide range of applications from the flight and propulsion systems of commercial airliners to the cruise control present in many modern automobiles.

In most of the cases, control engineers utilize feedback when designing control systems. This is often accomplished using a PID controller system. For example, in an automobile with cruise control the vehicle's speed is continuously monitored and fed back to the system, which adjusts the motor's torque accordingly. Where there is regular feedback, control theory can be used to determine how the system responds to such feedback. In practically all such systems stability is important and control theory can help ensure stability is achieved.

Although feedback is an important aspect of control engineering, control engineers may also work on the control of systems without feedback. This is known as open loop control. A classic example of open loop control is a washing machine that runs through a pre-determined cycle without the use of sensors.

Control Engineering Education

At many universities, control engineering courses are taught in electrical and electronic engineering, mechatronics engineering, mechanical engineering, and aerospace engineering. In others, control engineering is connected to computer science, as most control techniques today are implemented through computers, often as embedded systems (as in the automotive field). The field of control within chemical engineering is often known as process control. It deals primarily with the control of variables in a chemical process in a plant. It is taught as part of the undergraduate curriculum of any chemical engineering program and employs many of the same principles in control engineering. Other engineering disciplines also overlap with control engineering as it can be applied to any system for which a suitable model can be derived. However, specialised control engineering departments do exist, for example, the Department of Automatic Control and Systems Engineering at the University of Sheffield and the Department of Systems Engineering at the United States Naval Academy.

Control engineering has diversified applications that include science, finance management, and even human behavior. Students of control engineering may start with a linear control system course dealing with the time and complex-s domain, which requires a thorough background in elementary mathematics and Laplace transform, called classical control theory. In linear control, the student does frequency and time domain analysis. Digital control and nonlinear control courses require Z transformation and algebra respectively, and could be said to complete a basic control education.

Recent Advancement

Originally, control engineering was all about continuous systems. Development of computer control tools posed a requirement of discrete control system engineering because the communications between the computer-based digital controller and the physical system are governed by a computer clock. The equivalent to Laplace transform in the discrete domain is the Z-transform. Today, many of the control systems are computer controlled and they consist of both digital and analog components.

Therefore, at the design stage either digital components are mapped into the continuous domain and the design is carried out in the continuous domain, or analog components are mapped into discrete domain and design is carried out there. The first of these two methods is more commonly encountered in practice because many industrial systems have many continuous systems components, including mechanical, fluid, biological and analog electrical components, with a few digital controllers.

Similarly, the design technique has progressed from paper-and-ruler based manual design to computer-aided design and now to computer-automated design or CAutoD which has been made possible by evolutionary computation. CAutoD can be applied not just to tuning a predefined control scheme, but also to controller structure optimisation, system identification and invention of novel control systems, based purely upon a performance requirement, independent of any specific control scheme.

Resilient Control Systems extends the traditional focus on addressing only plant disturbances to frameworks, architectures and methods that address multiple types of unexpected disturbance. In particular, adapting and transforming behaviors of the control system in response to malicious actors, abnormal failure modes, undesirable human action, etc. Development of resilience technologies require the involvement of multidisciplinary teams to holistically address the performance challenges.

Aeroelasticity

Aeroelasticity is the branch of physics and engineering that studies the interactions between the inertial, elastic, and aerodynamic forces that occur when an elastic body is exposed to a fluid flow. Although historical studies have been focused on aeronautical applications, recent research has found applications in fields such as energy harvesting and understanding snoring. The study of aeroelasticity may be broadly classified into two fields: static aeroelasticity, which deals with the static or steady response of an elastic body to a fluid flow; and dynamic aeroelasticity, which deals with the body's dynamic (typically vibrational) response. Aeroelasticity draws on the study of fluid mechanics, solid mechanics, structural dynamics and dynamical systems. The synthesis of aeroelasticity with thermodynamics is known as aerothermoelasticity, and its synthesis with control theory is known as aeroservoelasticity.

History

The 2nd failure of Samuel Langley's prototype plane on the Potomac has been attributed to aeroelastic effects (specifically, torsional divergence). Problems with torsional divergence plagued aircraft in the First World War, and were solved largely by trial-and-error and ad-hoc stiffening of the wing. In 1926, Hans Reissner published a theory of wing divergence, leading to much further theoretical research on the subject.

In the development of aeronautical engineering at Caltech, Theodore von Kármán started a course "Elasticity applied to Areonautics". After teaching for one term he passed it over to Ernest Edwin Sechler, who went on to develop aeroelasticity in that course and in publication of textbooks on the subject.

In 1947, Arthur Roderick Collar defined aeroelasticity as *"the study of the mutual interaction that takes place within the triangle of the inertial, elastic, and aerodynamic forces acting on structural members exposed to an airstream, and the influence of this study on design."*

Static Aeroelasticity

In an aeroplane, two significant static aeroelastic effects may occur. *Divergence* is a phenomenon in which the elastic twist of the wing suddenly becomes theoretically infinite, typically causing the wing to fail spectacularly. *Control reversal* is a phenomenon occurring only in wings with ailerons or other control surfaces, in which these control surfaces reverse their usual functionality (e.g., the rolling direction associated with a given aileron moment is reversed).

Divergence

Divergence occurs when a lifting surface deflects under aerodynamic load so as to increase the applied load, or move the load so that the twisting effect on the structure is increased. The increased load deflects the structure further, which eventually brings the structure to the diverge point. Divergence can be understood as a simple property of the differential equation(s) governing the wing deflection. For example, modelling the aeroplane wing as an isotropic Euler–Bernoulli beam, the uncoupled torsional equation of motion is:

$$GJ\frac{d^2\theta}{dy^2} = -M'$$

Where y is the spanwise dimension, θ is the elastic twist of the beam, GJ is the torsional stiffness of the beam, L is the beam length, and M' is the aerodynamic moment per unit length. Under a simple lift forcing theory the aerodynamic moment is of the form:

$$M' = CU^2(\theta + \alpha_0)$$

Where C is a coefficient, U is the free-stream fluid velocity, and α_0 is the initial angle of attack. This yields an ordinary differential equation of the form:

$$\frac{d^2\theta}{dy^2} + \lambda^2\theta = -\lambda^2\alpha_0$$

Where:

$$\lambda^2 = CU^2/(GJ)$$

The boundary conditions for a clamped-free beam (i.e., a cantilever wing) are:

$$\theta\big|_{y=0} = \frac{d\theta}{dy}\bigg|_{y=L} = 0$$

Which yields the solution:

$$\theta = \alpha_0[\tan(\lambda L)\sin(\lambda y) + \cos(\lambda y) - 1]$$

As can be seen, for $\lambda L = \pi/2 + n\pi$, with arbitrary integer number n, $tan(\lambda L)$ is infinite. $n = 0$ corresponds to the point of torsional divergence. For given structural parameters, this will correspond to a single value of free-stream velocity U. This is the torsional diverengence speed. Note that for some special boundary conditions that may be implemented in a wind tunnel test of an airfoil (e.g., a torsional restraint positioned forward of the centre of lift) it is possible to eliminate the phenomenon of divergence altogether.

Control Reversal

Control surface reversal is the loss (or reversal) of the expected response of a control surface, due to deformation of the main lifting surface. For simple models (e.g. single aileron on an Euler-Benouilli beam), control reversal speeds can be derived analytically as for torsional divergence. Control reversal can be used to an aerodynamic advantage, and forms part of the Kaman servo-flap rotor design.

Dynamic Aeroelasticity

Dynamic Aeroelasticity studies the interactions among aerodynamic, elastic, and inertial forces. Examples of dynamic aeroelastic phenomena are:

Flutter

Flutter is a dynamic instability of an elastic structure in a fluid flow, caused by positive feedback between the body's deflection and the force exerted by the fluid flow. In a linear system, 'flutter point' is the point at which the structure is undergoing simple harmonic motion - zero net damping - and so any further decrease in net damping will result in a self-oscillation and eventual failure. 'Net damping' can be understood as the sum of the structure's natural positive damping, and the negative damping of the aerodynamic force. Flutter can be classified into two types: *hard flutter*, in which the net damping decreases very suddenly, very close to the flutter point; and *soft flutter*, in which the net damping decreases gradually. Methods of predicting flutter in linear structures include the p-method, the k-method and the p-k method. In water the mass ratio of the pitch inertia of the foil vs that of the circumscribing cylinder of fluid is generally too low for binary flutter to occur, as shown by explicit solution of the simplest pitch and heave flutter stability determinant.

For nonlinear systems, flutter is usually interpreted as a limit cycle oscillation (LCO), and methods from the study of dynamical systems can be used to determine the speed at which flutter will occur.

Structures exposed to aerodynamic forces — including wings and aerofoils, but also chimneys and bridges — are designed carefully within known parameters to avoid flutter. In complex structures where both the aerodynamics and the mechanical properties of the structure are not fully understood, flutter can be discounted only through detailed testing. Even changing the mass distribution of an aircraft or the stiffness of one component can induce flutter in an apparently unrelated aerodynamic component. At its mildest this can appear as a "buzz" in the aircraft structure, but at its most violent it can develop uncontrollably with great speed and cause serious damage to or lead to the destruction of the aircraft, as in Braniff Flight 542.

In some cases, automatic control systems have been demonstrated to help prevent or limit flutter-related structural vibration.

Flutter can also occur on structures other than aircraft. One famous example of flutter phenomena is the collapse of the original Tacoma Narrows Bridge.

Flutter as a controlled aerodynamic instability phenomenon is used intentionally and positively in windmills for generating electricity and in other works like making musical tones on ground-mounted devices, as well as on musical kites. Flutter is not always a destructive force; recent progress has been made in windmills for underserved communities in developing countries, designed specifically to take advantage of this effect. The oscillating motion allows variable-stroke waterpumping to match the variable power in the wind. Semirotary binary flutter can also have an upper critical airspeed at which it stops, affording automatic high wind protection The resulting Wing'd Pump has been designed to mount on the well it pumps or float on the pond it draws from. At its large scale the flutter is coupled by static gravity imbalance as well as dynamic imbalance. Further a gravity pendulum achieves large amplitude elasticity most practically. The same annual output can be achieved with wing length equal to a multiblade rotary windpump's diameter, in half the windspeed regime. P. Sharp and J. Hare showed a toy linear generator run by two flutter wings. Recently, researchers have demonstrated the ability to harvest energy directly from beam's self-induced, self-sustaining limit cycle oscillations in airflow of a flexible, piezoelectric beam placed in a wind tunnel. While the general approach to harvesting energy from these "aeroelastic" vibrations is to attach the beam to a secondary vibrating structure, such as a wing section, the new design eliminates the need for the secondary vibrating structure because the beam is designed so that it produces self-induced and self-sustaining vibrations (LCO). As a result, the new system can be made very small, which increases its efficiency and makes it more practical for applications, such as self-powered sensors.

The word *flutter* is typically linked to the form of aerodynamic instability discussed above. Though, a connection between dry friction and flutter instability in a simple mechanical system has been discovered, watch the movie for more details.

Buffeting

Buffeting is a high-frequency instability, caused by airflow separation or shock wave oscillations from one object striking another. It is caused by a sudden impulse of load increasing. It is a random forced vibration. Generally it affects the tail unit of the aircraft structure due to air flow downstream of the wing.

The methods for buffet detection are:

1. Pressure coefficient diagram

2. Pressure divergence at trailing edge

3. Computing separation from trailing edge based on Mach number

4. Normal force fluctuating divergence

Transonic Aeroelasticity

Flow is highly non-linear in the transonic regime, dominated by moving shock waves. It is mission-critical for aircraft that fly through transonic Mach numbers. The role of shock waves was first analyzed by Holt Ashley. A phenomenon that impacts stability of aircraft known as 'transonic dip', in which the flutter speed can get close to flight speed, was reported in May 1976 by Farmer and Hanson. of the Langley Research Center.

Prediction and Cure

Aeroelasticity involves not just the external aerodynamic loads and the way they change but also the structural, damping and mass characteristics of the aircraft. Prediction involves making a mathematical model of the aircraft as a series of masses connected by springs and dampers which are tuned to represent the dynamic characteristics of the aircraft structure. The model also includes details of applied aerodynamic forces and how they vary.

The model can be used to predict the flutter margin and, if necessary, test fixes to potential problems. Small carefully chosen changes to mass distribution and local structural stiffness can be very effective in solving aeroelastic problems.

Media

These videos detail the Active Aeroelastic Wing two-phase NASA-Air Force flight research program to investigate the potential of aerodynamically twisting flexible wings to improve maneuverability of high-performance aircraft at transonic and supersonic speeds, with traditional control surfaces such as ailerons and leading-edge flaps used to induce the twist.

Reaction Engine

A reaction engine is an engine or motor which provides thrust by expelling reaction mass, in accordance with Newton's third law of motion. This law of motion is most commonly paraphrased as: "For every action force there is an equal, but opposite, reaction force".

Examples include both jet engines and rocket engines, and more uncommon variations such as Hall effect thrusters, ion drives, mass drivers and nuclear pulse propulsion.

Energy Use

Propulsive Efficiency

For all reaction engines which carry their propellant onboard prior to use (such as rocket engines and electric propulsion drives) some energy must go into accelerating the reaction mass. Every engine will waste some energy, but even assuming 100% efficiency, the engine will need energy amounting to

$$\frac{1}{2}MV_e^2$$

(where M is the mass of propellent expended and V_e is the exhaust velocity), which is simply the energy to accelerate the exhaust.

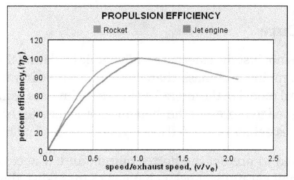

Due to energy carried away in the exhaust the energy efficiency of a reaction engine varies with the speed of the exhaust relative to the speed of the vehicle, this is called propulsive efficiency, blue is the curve for rocket-like reaction engines, red is for air-breathing (duct) reaction engines

Comparing the rocket equation (which shows how much energy ends up in the final vehicle) and the above equation (which shows the total energy required) shows that even with 100% engine efficiency, certainly not all energy supplied ends up in the vehicle - some of it, indeed usually most of it, ends up as kinetic energy of the exhaust.

Interestingly, if the specific impulse (I_{sp}) is fixed, for a mission delta-v, there is a particular I_{sp} that minimises the overall energy used by the rocket. This comes to an exhaust velocity of about ⅔ of the mission delta-v. Drives with a specific impulse that is both high and fixed such as Ion thrusters have exhaust velocities that can be enormously higher than this ideal, and thus end up powersource limited and give very low thrust. Where the vehicle performance is power limited, e.g. if solar power or nuclear power is used, then in the case of a large v_e the maximum acceleration is inversely proportional to it. Hence the time to reach a required delta-v is proportional to v_e. Thus the latter should not be too large.

On the other hand, if the exhaust velocity can be made to vary so that at each instant it is equal and opposite to the vehicle velocity then the absolute minimum energy usage is achieved. When this is achieved, the exhaust stops in space ^ and has no kinetic energy; and the propulsive efficiency is 100% all the energy ends up in the vehicle (in principle such a drive would be 100% efficient, in practice there would be thermal losses from within the drive system and residual heat in the exhaust). However, in most cases this uses an impractical quantity of propellant, but is a useful theoretical consideration.

Some drives (such as VASIMR or electrodeless plasma thruster) actually can significantly vary their exhaust velocity. This can help reduce propellant usage and improve acceleration at different stages of the flight. However the best energetic performance and acceleration is still obtained when the exhaust velocity is close to the vehicle speed. Proposed ion and plasma drives usually have exhaust velocities enormously higher than that ideal (in the case of VASIMR the lowest quoted speed is around 15 km/s compared to a mission delta-v from high Earth orbit to Mars of about 4 km/s).

For a mission, for example, when launching from or landing on a planet, the effects of gravitational attraction and any atmospheric drag must be overcome by using fuel. It is typical to combine the effects of these and other effects into an effective mission delta-v. For example, a launch mission to low Earth orbit requires about 9.3–10 km/s delta-v. These mission delta-vs are typically numerically integrated on a computer.

Cycle Efficiency

All reaction engines lose some energy, mostly as heat.

Different reaction engines have different efficiencies and losses. For example, rocket engines can be up to 60-70% energy efficient in terms of accelerating the propellant. The rest is lost as heat and thermal radiation, primarily in the exhaust.

Oberth Effect

Reaction engines are more energy efficient when they emit their reaction mass when the vehicle is travelling at high speed.

This is because the useful mechanical energy generated is simply force times distance, and when a thrust force is generated while the vehicle moves, then:

$$E = F \times d$$

where F is the force and d is the distance moved.

Dividing by length of time of motion we get:

$$\frac{E}{t} = P = \frac{F \times d}{t} = F \times v$$

Hence:

$$P = F \times v$$

where P is the useful power and v is the speed.

Hence you want v to be as high as possible; and a stationary engine does no useful work.

Types of Reaction Engines

- Rocket-like
  - o Rocket engine
  - o Electric propulsion, including VASIMR
- Airbreathing
  - o Turbojet
  - o Turbofan
  - o Pulsejet
  - o Ramjet
  - o Scramjet
- Liquid
  - o Pump-jet
- Rotary
  - o Aeolipile
- Solid exhaust
  - o Mass driver

Jet Engine

A Pratt & Whitney F100 turbofan engine for the F-15 Eagle being tested in the hush house at Florida Air National Guard base. The tunnel behind the engine muffles noise and allows exhaust to escape

A jet engine is a reaction engine discharging a fast-moving jet that generates thrust by jet propulsion. This broad definition includes turbojets, turbofans, rocket engines, ramjets, and pulse jets. In general, jet engines are combustion engines.

U.S. Air Force F-15E Strike Eagles

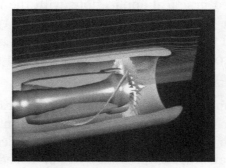
Simulation of a low-bypass turbofan's airflow.

Jet engine airflow during take-off.

In common parlance, the term *jet engine* loosely refers to an internal combustion air-breathing jet engine. These typically feature a rotating air compressor powered by a turbine, with the leftover power providing thrust via a propelling nozzle — this process is known as the Brayton thermodynamic cycle. Jet aircraft use such engines for long-distance travel. Early jet aircraft used turbojet engines which were relatively inefficient for subsonic flight. Modern subsonic jet aircraft usually use more complex high-bypass turbofan engines. These engines offer high speed and greater fuel efficiency than piston and propeller aeroengines over long distances.

The thrust of a typical jetliner engine went from 5,000 lbf (22,000 N) (de Havilland Ghost turbojet) in the 1950s to 115,000 lbf (510,000 N) (General Electric GE90 turbofan) in the 1990s, and their reliability went from 40 in-flight shutdowns per 100,000 engine flight hours to less than one in the late 1990s. This, combined with greatly decreased fuel consumption, permitted routine transatlantic flight by twin-engined airliners by the turn of the century, where before a similar journey would have required multiple fuel stops.

History

Jet engines date back to the invention of the aeolipile before the first century AD. This device directed steam power through two nozzles to cause a sphere to spin rapidly on its axis. So far as is known, it did not supply mechanical power and the potential practical applications of this invention did not receive recognition. Instead, it was seen as a curiosity.

Jet propulsion only gained practical applications with the invention of the gunpowder-powered rocket by the Chinese in the 13th century as a type of firework, and gradually progressed to propel formidable weaponry. However, although very powerful, at reasonable flight speeds rockets are very inefficient and so jet propulsion technology stalled for hundreds of years.

The earliest attempts at airbreathing jet engines were hybrid designs in which an external power source first compressed air, which was then mixed with fuel and burned for jet thrust. In one such system, called a *thermojet* by Secondo Campini but more commonly, motorjet, the air was compressed by a fan driven by a conventional piston engine. Examples of this type of design were the Caproni Campini N.1, and the Japanese Tsu-11 engine intended to power Ohka kamikaze planes towards the end of World War II. None were entirely successful and the N.1 ended up being slower than the same design with a traditional engine and propeller combination.

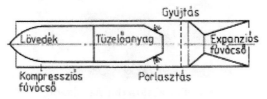

Albert Fonó's ramjet-cannonball from 1915

Even before the start of World War II, engineers were beginning to realize that engines driving propellers were self-limiting in terms of the maximum performance which could be attained; the limit was due to issues related to propeller efficiency, which declined as blade tips approached the speed of sound. If aircraft performance were ever to increase beyond such a barrier, a way would have to be found to use a different propulsion mechanism. This was the motivation behind the development of the gas turbine engine, commonly called a "jet" engine.

The key to a practical jet engine was the gas turbine, used to extract energy from the engine itself to drive the compressor. The gas turbine was not an idea developed in the

1930s: the patent for a stationary turbine was granted to John Barber in England in 1791. The first gas turbine to successfully run self-sustaining was built in 1903 by Norwegian engineer Ægidius Elling. Limitations in design and practical engineering and metallurgy prevented such engines reaching manufacture. The main problems were safety, reliability, weight and, especially, sustained operation.

The first patent for using a gas turbine to power an aircraft was filed in 1921 by Frenchman Maxime Guillaume. His engine was an axial-flow turbojet. Alan Arnold Griffith published *An Aerodynamic Theory of Turbine Design* in 1926 leading to experimental work at the RAE.

The Whittle W.2/700 engine flew in the Gloster E.28/39, the first British aircraft to fly with a turbojet engine, and the Gloster Meteor

In 1928, RAF College Cranwell cadet Frank Whittle formally submitted his ideas for a turbojet to his superiors. In October 1929 he developed his ideas further. On 16 January 1930 in England, Whittle submitted his first patent (granted in 1932). The patent showed a two-stage axial compressor feeding a single-sided centrifugal compressor. Practical axial compressors were made possible by ideas from A.A.Griffith in a seminal paper in 1926 ("An Aerodynamic Theory of Turbine Design"). Whittle would later concentrate on the simpler centrifugal compressor only, for a variety of practical reasons. Whittle had his first engine running in April 1937. It was liquid-fuelled, and included a self-contained fuel pump. Whittle's team experienced near-panic when the engine would not stop, accelerating even after the fuel was switched off. It turned out that fuel had leaked into the engine and accumulated in pools, so the engine would not stop until all the leaked fuel had burned off. Whittle was unable to interest the government in his invention, and development continued at a slow pace.

Heinkel He 178, the world's first aircraft to fly purely on turbojet power

In 1935 Hans von Ohain started work on a similar design in Germany, initially unaware of Whittle's work.

Von Ohain's first device was strictly experimental and could run only under external power, but he was able to demonstrate the basic concept. Ohain was then introduced to Ernst Heinkel, one of the larger aircraft industrialists of the day, who immediately saw the promise of the design. Heinkel had recently purchased the Hirth engine company, and Ohain and his master machinist Max Hahn were set up there as a new division of the Hirth company. They had their first HeS 1 centrifugal engine running by September 1937. Unlike Whittle's design, Ohain used hydrogen as fuel, supplied under external pressure. Their subsequent designs culminated in the gasoline-fuelled HeS 3 of 5 kN (1,100 lbf), which was fitted to Heinkel's simple and compact He 178 airframe and flown by Erich Warsitz in the early morning of August 27, 1939, from Rostock-Marienehe aerodrome, an impressively short time for development. The He 178 was the world's first jet plane.

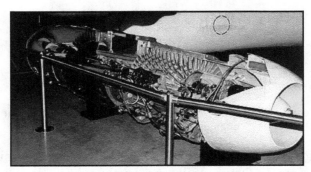

A cutaway of the Junkers Jumo 004 engine

Austrian Anselm Franz of Junkers' engine division (*Junkers Motoren* or "Jumo") introduced the axial-flow compressor in their jet engine. Jumo was assigned the next engine number in the RLM 109-0xx numbering sequence for gas turbine aircraft powerplants, "004", and the result was the Jumo 004 engine. After many lesser technical difficulties were solved, mass production of this engine started in 1944 as a powerplant for the world's first jet-fighter aircraft, the Messerschmitt Me 262 (and later the world's first jet-bomber aircraft, the Arado Ar 234). A variety of reasons conspired to delay the engine's availability, causing the fighter to arrive too late to improve Germany's position in World War II, however this was the first jet engine to be used in service.

Meanwhile, in Britain the Gloster E28/39 had its maiden flight on 15 May 1941 and the Gloster Meteor finally entered service with the RAF in July 1944. These were powered by turbojet engines from Power Jets Ltd., set up by Frank Whittle.

Following the end of the war the German jet aircraft and jet engines were extensively studied by the victorious allies and contributed to work on early Soviet and US jet fighters. The legacy of the axial-flow engine is seen in the fact that practically all jet engines on fixed-wing aircraft have had some inspiration from this design.

By the 1950s the jet engine was almost universal in combat aircraft, with the exception of cargo, liaison and other specialty types. By this point some of the British designs were already cleared for civilian use, and had appeared on early models like the de Havilland Comet and Avro Canada Jetliner. By the 1960s all large civilian aircraft were also jet powered, leaving the piston engine in low-cost niche roles such as cargo flights.

The efficiency of turbojet engines was still rather worse than piston engines, but by the 1970s, with the advent of high-bypass turbofan jet engines (an innovation not foreseen by the early commentators such as Edgar Buckingham, at high speeds and high altitudes that seemed absurd to them), fuel efficiency was about the same as the best piston and propeller engines.

Uses

A JT9D turbofan jet engine installed on a Boeing 747 aircraft.

Jet engines power jet aircraft, cruise missiles and unmanned aerial vehicles. In the form of rocket engines they power fireworks, model rocketry, spaceflight, and military missiles.

Jet engines have propelled high speed cars, particularly drag racers, with the all-time record held by a rocket car. A turbofan powered car, ThrustSSC, currently holds the land speed record.

Jet engine designs are frequently modified for non-aircraft applications, as industrial gas turbines or marine powerplants. These are used in electrical power generation, for powering water, natural gas, or oil pumps, and providing propulsion for ships and locomotives. Industrial gas turbines can create up to 50,000 shaft horsepower. Many of these engines are derived from older military turbojets such as the Pratt & Whitney J57 and J75 models. There is also a derivative of the P&W JT8D low-bypass turbofan that creates up to 35,000 HP.

Jet engines are also sometimes developed into, or share certain components such as engine cores, with turboshaft and turboprop engines, which are forms of gas turbine engines that are typically used to power helicopters and some propeller-driven aircraft..

Types

There are a large number of different types of jet engines, all of which achieve forward thrust from the principle of *jet propulsion.*

Airbreathing

Commonly aircraft are propelled by airbreathing jet engines. Most airbreathing jet engines that are in use are turbofan jet engines, which give good efficiency at speeds just below the speed of sound.

Turbine Powered

Gas turbines are rotary engines that extract energy from a flow of combustion gas. They have an upstream compressor coupled to a downstream turbine with a combustion chamber in-between. In aircraft engines, those three core components are often called the "gas generator." There are many different variations of gas turbines, but they all use a gas generator system of some type.

Turbojet

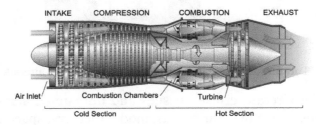

Turbojet engine

A turbojet engine is a gas turbine engine that works by compressing air with an inlet and a compressor (axial, centrifugal, or both), mixing fuel with the compressed air, burning the mixture in the combustor, and then passing the hot, high pressure air through a turbine and a nozzle. The compressor is powered by the turbine, which extracts energy from the expanding gas passing through it. The engine converts internal energy in the fuel to kinetic energy in the exhaust, producing thrust. All the air ingested by the inlet is passed through the compressor, combustor, and turbine, unlike the turbofan engine described below.

Turbofan

Turbofans differ from turbojets in that they have an additional fan at the front of the engine, which accelerates air in a duct bypassing the core gas turbine engine. Compared to a turbojet of identical thrust, a turbofan has a much larger air mass flow rate. Turbofans are the dominant engine type for medium and long-range airliners.

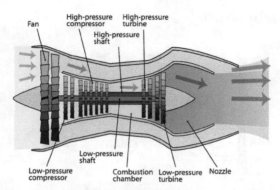

Schematic diagram illustrating the operation of a low-bypass turbofan engine.

The comparatively large frontal fan has several effects. The main effect is that the output of the engine as a whole has a much higher mass per second, and therefore generates much more thrust, despite not having ignited much of its airflow. Also, because the additional air has not been ignited, no extra fuel is needed to provide this thrust. The slower average velocity of the mixed exhaust air (low specific thrust) is also less wasteful of energy for subsonic flight, and allows the engine to be more efficient and much quieter, while the fan also allows greater thrust to be available at slow speeds. Together, the thrust produced by the fan and core are much more fuel efficient, and provides a much higher output, than could be produced by the core alone.

Turbofans are usually more efficient than turbojets at subsonic speeds, but their large frontal area also generates more drag. Therefore, in supersonic flight, and in military and other aircraft where absolute performance, weight, and drag have a higher priority than fuel efficiency, engines tend to either have smaller fans (or multiple smaller fans) or use other engine designs entirely. Typically, turbofans in civilian aircraft usually have a pronounced large front area to accommodate a very large fan, as their design involves a much larger mass of air bypassing the core so they can benefit from these effects, while in military aircraft, where noise and efficiency are less important compared to performance and drag, a smaller amount of air typically bypasses the core. Turbofans designed for subsonic civilian aircraft also usually have a just a single front fan, because their additional thrust is generated by a large additional mass of air which is only moderately compressed, rather than a smaller amount of air which is greatly compressed.

Because of these distinctions, turbofan engine designs are often categorized as low-bypass or high-bypass, depending upon the amount of air which bypasses the core of the engine. Low-bypass turbofans have a bypass ratio of around 2:1 or less, meaning that for each kilogram of air that passes through the core of the engine, two kilograms or less of air bypass the core. Low-bypass turbofans often use a mixed exhaust nozzle meaning that the bypassed flow and the core flow exit from the same nozzle. High-bypass turbofans often have ratios from 4:1 up to 8:1, with the Rolls-Royce Trent XWB approaching 10:1.

Turboprop and Turboshaft

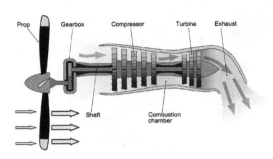

Turboprop engine

Turboprop engines are jet engine derivatives, still gas turbines, that extract work from the hot-exhaust jet to turn a rotating shaft, which is then used to produce thrust by some other means. While not strictly jet engines in that they rely on an auxiliary mechanism to produce thrust, turboprops are very similar to other turbine-based jet engines, and are often described as such.

In turboprop engines, a portion of the engine's thrust is produced by spinning a propeller, rather than relying solely on high-speed jet exhaust. As their jet thrust is augmented by a propeller, turboprops are occasionally referred to as a type of hybrid jet engine. They are quite similar to turbofans in many respects, except that they use a traditional propeller to provide the majority of thrust, rather than a ducted fan. Both fans and propellers are powered the same way, although most turboprops use gear-reduction between the turbine and the propeller (geared turbofans also feature gear reduction). While many turboprops generate the majority of their thrust with the propeller, the hot-jet exhaust is an important design point, and maximum thrust is obtained by matching thrust contributions of the propeller to the hot jet. Turboprops generally have better performance than turbojets or turbofans at low speeds where propeller efficiency is high, but become increasingly noisy and inefficient at high speeds.

Turboshaft engines are very similar to turboprops, differing in that nearly all energy in the exhaust is extracted to spin the rotating shaft, which is used to power machinery rather than a propeller, they therefore generate little to no jet thrust and are often used to power helicopters.

Propfan

A propfan engine (also called "unducted fan", "open rotor", or "ultra-high bypass") is a jet engine that uses its gas generator to power an exposed fan, similar to turboprop engines. Like turboprop engines, propfans generate most of their thrust from the propeller and not the exhaust jet. The primary difference between turboprop and propfan design is that the propeller blades on a propfan are highly swept to allow them to operate at speeds around Mach 0.8, which is competitive with modern commercial turbofans. These engines have the fuel efficiency advantages of turboprops with the performance capability of commercial turbofans. While significant research and testing (including flight testing) has been conducted on propfans, no propfan engines have entered production.

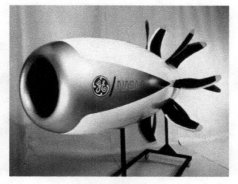

A propfan engine

Ram Powered

Ram powered jet engines are airbreathing engines similar to gas turbine engines and they both follow the Brayton cycle. Gas turbine and ram powered engines differ, however, in how they compress the incoming airflow. Whereas gas turbine engines use axial or centrifugal compressors to compress incoming air, ram engines rely only on air compressed through the inlet or diffuser. Ram powered engines are considered the most simple type of air breathing jet engine because they can contain no moving parts.

Ramjet

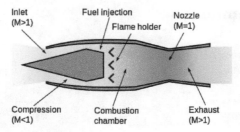

A schematic of a ramjet engine, where "M" is the Mach number of the airflow.

Ramjets are the most basic type of ram powered jet engines. They consist of three sections; an inlet to compress incoming air, a combustor to inject and combust fuel, and a nozzle to expel the hot gases and produce thrust. Ramjets require a relatively high speed to efficiently compress the incoming air, so ramjets cannot operate at a standstill and they are most efficient at supersonic speeds. A key trait of ramjet engines is that combustion is done at subsonic speeds. The supersonic incoming air is dramatically slowed through the inlet, where it is then combusted at the much slower, subsonic, speeds. The faster the incoming air is, however, the less efficient it becomes to slow it to subsonic speeds. Therefore, ramjet engines are limited to approximately Mach 5.

Scramjet

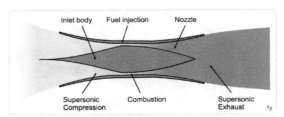

Scramjet engine operation

Scramjets are mechanically very similar to ramjets. Like a ramjet, they consist of an inlet, a combustor, and a nozzle. The primary difference between ramjets and scramjets is that scramjets do not slow the oncoming airflow to subsonic speeds for combustion, they use supersonic combustion instead. The name "scramjet" comes from "Supersonic Combusting Ramjet." Since scramjets use supersonic combustion they can operate at speeds above Mach 6 where traditional ramjets are too inefficient. Another difference between ramjets and scramjets comes from how each type of engine compresses the oncoming airflow: while the inlet provides most of the compression for ramjets, the high speeds at which scramjets operate allow them to take advantage of the compression generated by shock waves, primarily oblique shocks.

Very few scramjet engines have ever been built and flown. In May 2010 the Boeing X-51 set the endurance record for the longest scramjet burn at over 200 seconds.

Non-continuous Combustion

Rocket

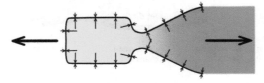

Rocket engine propulsion

The rocket engine uses the same basic physical principles as the jet engine for propulsion via thrust, but is distinct in that it does not require atmospheric air to provide oxygen; the rocket carries all components of the reaction mass. This allows them to operate at arbitrary altitudes and in space.

This type of engine is used for launching satellites, space exploration and manned access, and permitted landing on the moon in 1969.

Rocket engines are used for high altitude flights, or anywhere where very high accelerations are needed since rocket engines themselves have a very high thrust-to-weight ratio.

However, the high exhaust speed and the heavier, oxidizer-rich propellant results in far more propellant use than turbofans. Even so, at extremely high speeds they become energy-efficient.

An approximate equation for the net thrust of a rocket engine is:

$$F_N = \dot{m} g_0 I_{sp-vac} - A_e p$$

Where F_N is the net thrust, $I_{sp(vac)}$ is the specific impulse, g_0 is a standard gravity, $\dot{m}$ is the propellant flow in kg/s, A_e is the cross-sectional area at the exit of the exhaust nozzle, and p is the atmospheric pressure.

Hybrid

Combined cycle engines simultaneously use 2 or more different jet engine operating principles.

Water Jet

A water jet, or pump jet, is a marine propulsion system that utilizes a jet of water. The mechanical arrangement may be a ducted propeller with nozzle, or a centrifugal compressor and nozzle.

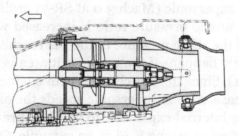

A pump jet schematic.

Type	Description	Advantages	Disadvantages
Water jet	For propelling water rockets and jetboats; squirts water out the back through a nozzle	In boats, can run in shallow water, high acceleration, no risk of engine overload (unlike propellers), less noise and vibration, highly maneuverable at all boat speeds, high speed efficiency, less vulnerable to damage from debris, very reliable, more load flexibility, less harmful to wildlife	Can be less efficient than a propeller at low speed, more expensive, higher weight in boat due to entrained water, will not perform well if boat is heavier than the jet is sized for

General Physical Principles

All jet engines are reaction engines that generate thrust by emitting a jet of fluid rearwards at relatively high speed. The forces on the inside of the engine needed to create this jet give a strong thrust on the engine which pushes the craft forwards.

Jet engines make their jet from propellant from tankage that is attached to the engine (as in a 'rocket') as well as in duct engines (those commonly used on aircraft) by ingesting an external fluid (very typically air) and expelling it at higher speed.

Propelling Nozzle

The propelling nozzle is the key component of all jet engines as it creates the exhaust jet. Propelling nozzles turn internal and pressure energy into high velocity kinetic energy. The total pressure and temperature don't change through the nozzle but their static values drop as the gas speeds up.

The velocity of the air entering the nozzle is low, about Mach 0.4, a prerequisite for minimizing pressure losses in the duct leading to the nozzle. The temperature entering the nozzle may be as low as sea level ambient for a fan nozzle in the cold air at cruise altitudes. It may be as high as the 1000K exhaust gas temperature for a supersonic afterburning engine or 2200K with afterburner lit. The pressure entering the nozzle may vary from 1.5 times the pressure outside the nozzle, for a single stage fan, to 30 times for the fastest manned aircraft at mach 3+.

The velocity of the gas leaving a convergent nozzle may be subsonic or sonic (Mach 1) at low flight speeds or supersonic (Mach 3.0 at SR-71 cruise) for a con-di nozzle at higher speeds where the nozzle pressure ratio is increased with the intake ram. The nozzle thrust is highest if the static pressure of the gas reaches the ambient value as it leaves the nozzle. This only happens if the nozzle exit area is the correct value for the nozzle pressure ratio (npr). Since the npr changes with engine thrust setting and flight speed this is seldom the case. Also at supersonic speeds the divergent area is less than required to give complete internal expansion to ambient pressure as a trade-off with external body drag. Whitford gives the F-16 as an example. Other underexpanded examples were the XB-70 and SR-71.

The nozzle size, together with the area of the turbine nozzles, determines the operating pressure of the compressor.

Thrust

Origin of Engine Thrust

The familiar explanation for jet thrust is a "black box" description which only looks at what goes in to the engine, air and fuel, and what comes out, exhaust gas and an unbalanced force. This force, called thrust, is the sum of the momentum difference between entry and exit and any unbalanced pressure force between entry and exit, as explained in "Thrust calculation". As an example, an early turbojet, the Bristol Olympus Mk. 101, had a momentum thrust of 9300 lb. and a pressure thrust of 1800 lb. giving a total of 11,100 lb. Looking inside the "black box" shows that the thrust results from all the unbalanced momentum and pressure forces created within the engine itself. These forces, some forwards and some rearwards, are across all the internal parts, both stationary and rotating, such as ducts, compressors, etc., which are in the primary gas flow which flows through the engine from front to rear. The algebraic sum of all these forces is delivered to the airframe for propulsion. "Flight" gives examples of these internal forces for two early jet engines, the Rolls-Royce Avon Ra.14 and the de Havilland Goblin

Transferring Thrust to the Aircraft

The engine thrust acts along the engine centreline. The aircraft "holds" the engine on the outer casing of the engine at some distance from the engine centreline (at the engine mounts). This arrangement causes the engine casing to bend (known as backbone bending) and the round rotor casings to distort (ovalization). Distortion of the engine structure has to be controlled with suitable mount locations to maintain acceptable rotor and seal clearances and prevent rubbing. A well-publicized example of excessive structural deformation occurred with the original Pratt & Whitney JT9D engine installation in the Boeing 747 aircraft. The engine mounting arrangement had to be revised with the addition of an extra thrust frame to reduce the casing deflections to an acceptable amount.

Rotor Thrust

The rotor thrust on a thrust bearing is not related to the engine thrust. It may even change direction at some RPM. The bearing load is determined by bearing life considerations. Although the aerodynamic loads on the compressor and turbine blades contribute to the rotor thrust they are small compared to cavity loads inside the rotor which result from the secondary air system pressures and sealing diameters on discs, etc. To keep the load within the bearing specification seal diameters are chosen accordingly as, many years ago, on the backface of the impeller in the de Havilland Ghost engine. Sometimes an extra disc known as a balance piston has to be added inside the rotor. An early turbojet example with a balance piston was the Rolls-Royce Avon.

Thrust Calculation

The net thrust (F_N) of a turbojet is given by:

$$F_N = (\dot{m}_{air} + \dot{m}_{fuel})v_e - \dot{m}_{air}v$$

where:

$\dot{m}_{air}$ = the mass rate of air flow through the engine

$\dot{m}_{fuel}$ = the mass rate of fuel flow entering the engine

v_e = the velocity of the jet (the exhaust plume) and is assumed to be less than sonic velocity

v = the velocity of the air intake = the true airspeed of the aircraft

$(\dot{m}_{air} + \dot{m}_{fuel})v_e$ = the nozzle gross thrust (F_G)

$\dot{m}_{air}\,v$ = the ram drag of the intake air

The above equation applies only for air-breathing jet engines. It does not apply to rocket engines. Most types of jet engine have an air intake, which provides the bulk of the fluid exiting the exhaust. Conventional rocket engines, however, do not have an intake, the oxidizer and fuel both being carried within the vehicle. Therefore, rocket engines do not have ram drag and the gross thrust of the rocket engine nozzle is the net thrust of the engine. Consequently, the thrust characteristics of a rocket motor are different from that of an air breathing jet engine, and thrust is independent of velocity.

If the velocity of the jet from a jet engine is equal to sonic velocity, the jet engine's nozzle is said to be choked. If the nozzle is choked, the pressure at the nozzle exit plane is greater than atmospheric pressure, and extra terms must be added to the above equation to account for the pressure thrust.

The rate of flow of fuel entering the engine is very small compared with the rate of flow of air. If the contribution of fuel to the nozzle gross thrust is ignored, the net thrust is:

$$F_N = \dot{m}_{air}(v_e - v)$$

The velocity of the jet (v_e) must exceed the true airspeed of the aircraft (v) if there is to be a net forward thrust on the aircraft. The velocity (v_e) can be calculated thermodynamically based on adiabatic expansion.

Thrust Augmentation

Thrust augmentation has taken many forms, most commonly to supplement inadequate take-off thrust. Some early jet aircraft needed rocket assistance to take off from high altitude airfields or when the day temperature was high. A more recent aircraft, the Tupolev Tu-22 supersonic bomber, was fitted with four SPRD-63 boosters for take-off. Possibly the most extreme requirement needing rocket assistance, and which was short-lived, was zero-length launching. Almost as extreme, but very common, is catapult assistance from aircraft carriers. Rocket assistance has also been used during flight. The SEPR 841 booster engine was used on the Dassault Mirage for high altitude interception.

Early aft-fan arrangements which added bypass airflow to a turbojet were known as thrust augmentors. The aft-fan fitted to the General Electric CJ805-3 turbojet augmented the take-off thrust from 11,650lb to 16,100lb.

Water, or other coolant, injection into the compressor or combustion chamber and fuel injection into the jetpipe (afterburning/reheat) became standard ways to increase thrust, known as 'wet' thrust to differentiate with the no-augmentation 'dry' thrust.

Coolant injection (pre-compressor cooling) has been used, together with afterburning, to increase thrust at supersonic speeds. The 'Skyburner' McDonnell Douglas F-4 Phantom II set a world speed record using water injection in front of the engine.

At high Mach numbers afterburners supply progressively more of the engine thrust as the thrust from the turbomachine drops off towards zero at which speed the engine pressure ratio (epr) has fallen to 1.0 and all the engine thrust comes from the afterburner. The afterburner also has to make up for the pressure loss across the turbomachine which is a drag item at higher speeds where the epr will be less than 1.0.

Thrust augmentation of existing afterburning engine installations for special short-duration tasks has been the subject of studies for launching small payloads into low earth orbits using aircraft such as McDonnell Douglas F-4 Phantom II, McDonnell Douglas F-15 Eagle, Dassault Rafale and Mikoyan MiG-31, and also for carrying experimental packages to high altitudes using a Lockheed SR-71. In the first case an increase in the existing maximum speed capability is required for orbital launches. In the second case an increase in thrust within the existing speed capability is required. Compressor inlet cooling is used in the first case. A compressor map shows that the airflow reduces with increasing compressor inlet temperature although the compressor is still running at maximum RPM (but reduced aerodynamic speed). Compressor inlet cooling increases the aerodynamic speed and flow and thrust. In the second case a small increase in the maximum mechanical speed and turbine temperature were allowed, together with nitrous oxide injection into the afterburner and simultaneous increase in afterburner fuel flow.

Energy Efficiency Relating to Aircraft Jet Engines

This overview highlights where energy losses occur in complete jet aircraft powerplants or engine installations. It includes mention of inlet and exhaust nozzle losses which become increasingly significant at the high flight speeds achieved by some manned aircraft since only a small proportion, 17% for the SR-71 powerplant and 8% for the Concorde powerplant, of the thrust transmitted to the airframe came from the engine.

A jet engine at rest, as on a test stand, sucks in fuel and tries to thrust itself forward. How well it does this is judged by how much fuel it uses and what force is required to restrain it. This is a measure of its efficiency. If something deteriorates inside the engine (known as performance deterioration) it will be less efficient and this will show when the fuel produces less thrust. If a change is made to an internal part which allows the air/combustion gases to flow more smoothly the engine will be more efficient and use less fuel. A standard definition is used to assess how different things change engine efficiency and also to allow comparisons to be made between different engines. This definition is called specific fuel consumption, or how much fuel is needed to produce one unit of thrust. For example, it will be known for a particular engine design that if some bumps in a bypass duct are smoothed out the air will flow more smoothly giving a pressure loss reduction of x% and y% less fuel will be needed to get the take-off thrust, for example. This understanding comes under the engineering discipline Jet engine performance. How efficiency is affected by forward speed and by supplying energy to aircraft systems is mentioned later.

The efficiency of the engine is controlled primarily by the operating conditions inside the engine which are the pressure produced by the compressor and the temperature of the combustion gases at the first set of rotating turbine blades. The pressure is the highest air pressure in the engine. The turbine rotor temperature is not the highest in the engine but is the highest at which energy transfer takes place (higher temperatures occur in the combustor). The above pressure and temperature are shown on a Thermodynamic cycle diagram.

The efficiency is further modified by how smoothly the air and the combustion gases flow through the engine, how well the flow is aligned (known as incidence angle) with the moving and stationary passages in the compressors and turbines. Non-optimum angles, as well as non-optimum passage and blade shapes can cause thickening and separation of Boundary layers and formation of Shock waves as explained in Effects of Mach number and shock losses in turbomachines. It is important to slow the flow (lower speed means less pressure losses or Pressure drop) when it travels through ducts connecting the different parts. How well the individual components contribute to turning fuel into thrust is quantified by measures like efficiencies for the compressors, turbines and combustor and pressure losses for the ducts. These are shown as lines on a Thermodynamic cycle diagram.

The engine efficiency, or thermal efficiency, known as η_{th}. is dependent on the Thermodynamic cycle parameters, maximum pressure and temperature, and on component efficiencies, $\eta_{compressor}$, $\eta_{combustion}$ and $\eta_{turbine}$ and duct pressure losses.

The engine needs compressed air for itself just to run successfully. This air comes from its own compressor and is called secondary air. It does not contribute to making thrust so makes the engine less efficient. It is used to preserve the mechanical integrity of the engine, to stop parts overheating and to prevent oil escaping from bearings for example. Only some of this air taken from the compressors returns to the turbine flow to contribute to thrust production. Any reduction in the amount needed improves the engine efficiency. Again, it will be known for a particular engine design that a reduced requirement for cooling flow of x% will reduce the specific fuel consumption by y%. In other words, less fuel will be required to give take-off thrust, for example. The engine is more efficient.

All of the above considerations are basic to the engine running on its own and, at the same time, doing nothing useful, i.e. it is not moving an aircraft or supplying energy for the aircraft's electrical, hydraulic and air systems. In the aircraft the engine gives away some of its thrust-producing potential, or fuel, to power these systems. These requirements, which cause installation losses, reduce its efficiency. It is using some fuel that does not contribute to the engine's thrust.

Finally, when the aircraft is flying the propelling jet itself contains wasted kinetic energy after it has left the engine. This is quantified by the term propulsive, or Froude, efficiency η_p and may be reduced by redesigning the engine to give it bypass flow and a lower speed for the propelling jet, for example as a turboprop or turbofan engine. At the same time forward speed increases the η_{th} by increasing the Overall pressure ratio.

The overall efficiency of the engine at flight speed is defined as $\eta_o = \eta_p \eta_{th}$.

The η_o at flight speed depends on how well the intake compresses the air before it is handed over to the engine compressors. The intake compression ratio, which can be as high as 32:1 at Mach 3, adds to that of the engine compressor to give the Overall pressure ratio and η_{th} for the Thermodynamic cycle. How well it does this is defined by its pressure recovery or measure of the losses in the intake. Mach 3 manned flight has provided an interesting illustration of how these losses can increase dramatically in an instant. The North American XB-70 Valkyrie and Lockheed SR-71 Blackbird at Mach 3 each had pressure recoveries of about 0.8, due to relatively low losses during the compression process, i.e. through systems of multiple shocks. During an 'unstart' the efficient shock system would be replaced by a very inefficient single shock beyond the inlet and an intake pressure recovery of about 0.3 and a correspondingly low pressure ratio.

The propelling nozzle at speeds above about Mach 2 usually has extra internal thrust losses because the exit area is not big enough as a trade-off with external afterbody drag.

Although a bypass engine improves propulsive efficiency it incurs losses of its own inside the engine itself. Machinery has to be added to transfer energy from the gas generator to a bypass airflow. The low loss from the propelling nozzle of a turbojet is added to with extra losses due to inefficiencies in the added turbine and fan. These may be included in a transmission, or transfer, efficiency η_T. However, these losses are more than made up by the improvement in propulsive efficiency. There are also extra pressure losses in the bypass duct and an extra propelling nozzle.

With the advent of turbofans with their loss-making machinery what goes on inside the engine has been separated by Bennett, for example, between gas generator and transfer machinery giving $\eta_o = \eta_p \eta_{th} \eta_T$.

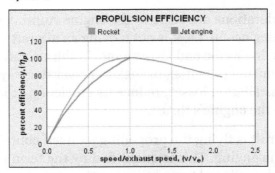

Dependence of propulsion efficiency (η) upon the vehicle speed/exhaust velocity ratio (v/v_e) for air-breathing jet and rocket engines.

The energy efficiency (η_o) of jet engines installed in vehicles has two main components:

- *propulsive efficiency* (η_p): how much of the energy of the jet ends up in the vehicle body rather than being carried away as kinetic energy of the jet.

- *cycle efficiency* (η_{th}): how efficiently the engine can accelerate the jet

Even though overall energy efficiency η_o is:

$$\eta_o = \eta_p \eta_{th}$$

for all jet engines the *propulsive efficiency* is highest as the exhaust jet velocity gets closer to the vehicle speed as this gives the smallest residual kinetic energy. For an air-breathing engine an exhaust velocity equal to the vehicle velocity, or a η_p equal to one, gives zero thrust with no net momentum change. The formula for air-breathing engines moving at speed v with an exhaust velocity v_e, and neglecting fuel flow, is:

$$\eta_p = \frac{2}{1+\dfrac{v_e}{v}}$$

And for a rocket:

$$\eta_p = \frac{2(\dfrac{v}{v_e})}{1+(\dfrac{v}{v_e})^2}$$

In addition to propulsive efficiency, another factor is *cycle efficiency*; a jet engine is a form of heat engine. Heat engine efficiency is determined by the ratio of temperatures reached in the engine to that exhausted at the nozzle. This has improved constantly over time as new materials have been introduced to allow higher maximum cycle temperatures. For example, composite materials, combining metals with ceramics, have been developed for HP turbine blades, which run at the maximum cycle temperature. The efficiency is also limited by the overall pressure ratio that can be achieved. Cycle efficiency is highest in rocket engines (~60+%), as they can achieve extremely high combustion temperatures. Cycle efficiency in turbojet and similar is nearer to 30%, due to much lower peak cycle temperatures.

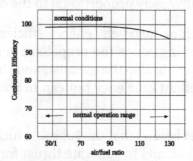

Typical combustion efficiency of an aircraft gas turbine over the operational range.

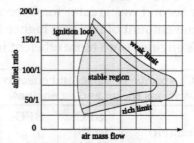

Typical combustion stability limits of an aircraft gas turbine.

The combustion efficiency of most aircraft gas turbine engines at sea level takeoff conditions is almost 100%. It decreases nonlinearly to 98% at altitude cruise conditions. Air-fuel ratio ranges from 50:1 to 130:1. For any type of combustion chamber there is a *rich* and *weak limit* to the air-fuel ratio, beyond which the flame is extinguished. The range of air-fuel ratio between the rich and weak limits is reduced with an increase of air velocity. If the increasing air mass flow reduces the fuel ratio below certain value, flame extinction occurs.

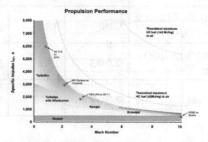

Specific impulse as a function of speed for different jet types with kerosene fuel (hydrogen I_{sp} would be about twice as high). Although efficiency plummets with speed, greater distances are covered. Efficiency per unit distance (per km or mile) is roughly independent of speed for jet engines as a group; however, airframes become inefficient at supersonic speeds.

Consumption of Fuel or Propellant

A closely related (but different) concept to energy efficiency is the rate of consumption of propellant mass. Propellant consumption in jet engines is measured by Specific Fuel Consumption, Specific impulse or Effective exhaust velocity. They all measure the same thing. Specific impulse and effective exhaust velocity are strictly proportional, whereas specific fuel consumption is inversely proportional to the others.

For airbreathing engines such as turbojets, energy efficiency and propellant (fuel) efficiency are much the same thing, since the propellant is a fuel and the source of energy. In rocketry, the propellant is also the exhaust, and this means that a high energy propellant gives better propellant efficiency but can in some cases actually give *lower* energy efficiency.

It can be seen in the table (just below) that the subsonic turbofans such as General Electric's CF6 turbofan use a lot less fuel to generate thrust for a second than did the Concorde's Rolls-Royce/Snecma Olympus 593 turbojet. However, since energy is force times distance and the distance per second was greater for Concorde, the actual power generated by the engine for the same amount of fuel was higher for Concorde at Mach 2 than the CF6. Thus, the Concorde's engines were more efficient in terms of energy per mile.

Specific fuel consumption (SFC), specific impulse, and effective exhaust velocity numbers for various rocket and jet engines.					
Engine type	**Scenario**	**SFC in lb/(lbf·h)**	**SFC in g/(kN·s)**	**Specific impulse (s)**	**Effective exhaust velocity (m/s)**
NK-33 rocket engine	Vacuum	10.9	308	331	3250
SSME rocket engine	Space shuttle vacuum	7.95	225	453	4440
Ramjet	Mach 1	4.5	130	800	7800
J-58 turbojet	SR-71 at Mach 3.2 (Wet)	1.9	54	1900	19000
Eurojet EJ200	Reheat	1.7	47	2200	21000
Rolls-Royce/Snecma Olympus 593 turbojet	Concorde Mach 2 cruise (Dry)	1.195	33.8	3010	29500
CF6-80C2B1F turbofan	Boeing 747-400 cruise	0.605	17.1	5950	58400
General Electric CF6 turbofan	Sea level	0.307	8.7	11700	115000

Thrust-to-weight Ratio

The thrust-to-weight ratio of jet engines with similar configurations varies with scale, but is mostly a function of engine construction technology. For a given engine, the lighter the engine, the better the thrust-to-weight is, the less fuel is used to compensate for drag due to the lift needed to carry the engine weight, or to accelerate the mass of the engine.

As can be seen in the following table, rocket engines generally achieve much higher thrust-to-weight ratios than duct engines such as turbojet and turbofan engines. This is primarily because rockets almost universally use dense liquid or solid reaction mass which gives a much smaller volume and hence the pressurization system that supplies the nozzle is much smaller and lighter for the same performance. Duct engines have to deal with air which is two to three orders of magnitude less dense and this gives pressures over much larger areas, which in turn results in more engineering materials being needed to hold the engine together and for the air compressor.

Rocket thrusts are vacuum thrusts unless otherwise noted

Comparison of Types

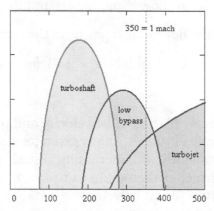

Comparative suitability for (left to right) turboshaft, low bypass and turbojet to fly at 10 km altitude in various speeds. Horizontal axis - speed, m/s. Vertical axis displays engine efficiency.

Propeller engines handle larger air mass flows, and give them smaller acceleration, than jet engines. Since the increase in air speed is small, at high flight speeds the thrust available to propeller-driven aeroplanes is small. However, at low speeds, these engines benefit from relatively high propulsive efficiency.

On the other hand, turbojets accelerate a much smaller mass flow of intake air and burned fuel, but they then reject it at very high speed. When a de Laval nozzle is used to accelerate a hot engine exhaust, the outlet velocity may be locally supersonic. Turbojets are particularly suitable for aircraft travelling at very high speeds.

Turbofans have a mixed exhaust consisting of the bypass air and the hot combustion product gas from the core engine. The amount of air that bypasses the core engine compared to the amount flowing into the engine determines what is called a turbofan's bypass ratio (BPR).

While a turbojet engine uses all of the engine's output to produce thrust in the form of a hot high-velocity exhaust gas jet, a turbofan's cool low-velocity bypass air yields between 30% and 70% of the total thrust produced by a turbofan system.

The net thrust (F_N) generated by a turbofan is:

$$F_N = \dot{m}_e v_e - \dot{m}_o v_o + BPR(\dot{m}_c v_f)$$

where:

$\dot{m}_e$ = the mass rate of hot combustion exhaust flow from the core engine

$\dot{m}_o$ = the mass rate of total air flow entering the turbofan = $\dot{m}_c + \dot{m}_f$

$\dot{m}_c$ = the mass rate of intake air that flows to the core engine

$\dot{m}_f$ = the mass rate of intake air that bypasses the core engine

V_f = the velocity of the air flow bypassed around the core engine

V_e = the velocity of the hot exhaust gas from the core engine

V_o = the velocity of the total air intake = the true airspeed of the aircraft

BPR = Bypass Ratio

Rocket engines have extremely high exhaust velocity and thus are best suited for high speeds (hypersonic) and great altitudes. At any given throttle, the thrust and efficiency of a rocket motor improves slightly with increasing altitude (because the back-pressure falls thus increasing net thrust at the nozzle exit plane), whereas with a turbojet (or turbofan) the falling density of the air entering the intake (and the hot gases leaving the nozzle) causes the net thrust to decrease with increasing altitude. Rocket engines are more efficient than even scramjets above roughly Mach 15.

Altitude and Speed

With the exception of scramjets, jet engines, deprived of their inlet systems can only accept air at around half the speed of sound. The inlet system's job for transonic and supersonic aircraft is to slow the air and perform some of the compression.

The limit on maximum altitude for engines is set by flammability- at very high altitudes the air becomes too thin to burn, or after compression, too hot. For turbojet engines altitudes of about 40 km appear to be possible, whereas for ramjet engines 55 km may be achievable. Scramjets may theoretically manage 75 km. Rocket engines of course have no upper limit.

At more modest altitudes, flying faster compresses the air at the front of the engine, and this greatly heats the air. The upper limit is usually thought to be about Mach 5-8, as above about Mach 5.5, the atmospheric nitrogen tends to react due to the high temperatures at the inlet and this consumes significant energy. The exception to this is scramjets which may be able to achieve about Mach 15 or more, as they avoid slowing the air, and rockets again have no particular speed limit.

Noise

The noise emitted by a jet engine has many sources. These include, in the case of gas turbine engines, the fan, compressor, combustor, turbine and propelling jet/s.

The propelling jet produces jet noise which is caused by the violent mixing action of the high speed jet with the surrounding air. In the subsonic case the noise is produced by eddies and in the supersonic case by Mach waves. The sound power radiated from a jet varies with the jet velocity raised to the eighth power for velocities up to 2,000 ft/sec and varies with the velocity cubed above 2,000 ft/sec. Thus, the lower speed exhaust jets emitted from engines such as high bypass turbofans are the quietest, whereas the fastest jets, such as rockets, turbojets, and ramjets, are the loudest. For commercial jet aircraft the jet noise has reduced from the turbojet through bypass engines to turbofans as a result of a progressive reduction in propelling jet velocities. For example, the JT8D, a bypass engine, has a jet velocity of 1450 ft/sec whereas the JT9D, a turbofan, has jet velocities of 885 ft/sec (cold) and 1190 ft/sec (hot).

The advent of the turbofan replaced the very distinctive jet noise with another sound known as "buzz saw" noise. The origin is the shockwaves originating at the supersonic fan blades at takeoff thrust.

Flight Dynamics (Spacecraft)

Spacecraft flight dynamics is the science of space vehicle performance, stability, and control. It requires analysis of the six degrees of freedom of the vehicle's flight, which are similar to those of aircraft: translation in three dimensional axes; and its orientation about the vehicle's center of mass in these axes, known as *pitch*, *roll* and *yaw*, with respect to a defined frame of reference.

Dynamics is the modeling of the changing position and orientation of a vehicle, in response to external forces acting on the body. For a spacecraft, these forces are of three types: propulsive force (usually provided by the vehicle's engine thrust); gravitational force exerted by the Earth or other celestial bodies; and aerodynamic lift and drag (when flying in the atmosphere of the Earth or other body, such as Mars or Venus). The vehicle's attitude must be taken into account because of its effect on the aerodynamic and propulsive forces. There are other reasons, unrelated to flight dynamics, for controlling the vehicle's attitude in non-powered flight (e.g., thermal control, solar power generation, communications, or astronomical observation).

The principles of flight dynamics are normally used to control a spacecraft by means of an inertial navigation system in conjunction with an attitude control system. Together, they create a subsystem of the spacecraft bus often called ADCS.

Basic Principles

A spacecraft's flight is determined by application of Newton's second law of motion:

$$\mathbf{F} = m\mathbf{a},$$

where $\mathbf{F}$ is the vector sum of all forces exerted on the vehicle, m is its current mass, and a is the acceleration vector, the instantaneous rate of change of velocity (v), which in turn is the instantaneous rate of change of displacement. Solving for a, acceleration equals the force sum divided by mass. Acceleration is integrated over time to get velocity, and velocity is in turn integrated to get position.

Aerodynamic forces, present near a body with significant atmosphere such as Earth, Mars or Venus, are analyzed as: lift, defined as the force component perpendicular to the direction of flight (not necessarily upward to balance gravity, as for an airplane); and drag, the component parallel to, and in the opposite direction of flight. Lift and drag are modeled as the products of a coefficient times dynamic pressure acting on a reference area:

$$\mathbf{L} = C_L q A_{ref}$$
$$\mathbf{D} = C_D q A_{ref}$$

where:

- C_L is roughly linear with α, the angle of attack between the vehicle axis and the direction of flight (up to a limiting value), and is 0 at $\alpha = 0$ for an axisymmetric body;

- C_D varies with α^2;

- C_L and C_D vary with Reynolds number and Mach number;

- q, the dynamic pressure, is equal to $1/2\ \rho\ v^2$, where ρ is atmospheric density, modeled for Earth as a function of altitude in the International Standard Atmosphere (using an assumed temperature distribution, hydrostatic pressure variation, and the ideal gas law); and

- A_{ref} is a characteristic area of the vehicle, such as cross-sectional area at the maximum diameter.

Powered Flight

Flight calculations are made quite precisely for space missions, taking into account such factors as the Earth's oblateness and non-uniform mass distribution; gravitational forces of all nearby bodies, including the Moon, Sun, and other planets; and three-dimensional flight path. For preliminary studies, some simplifying assumptions can be made (spherical, uniform planet; two-body patched conic approximation; and co-planar local flight path) with reasonably small loss of accuracy.

The general case of a launch from Earth must take engine thrust, aerodynamic forces and gravity into account. The acceleration equation can be reduced from vector to scalar form by resolving it into tangential and angular components. The two equations thus become:

$$\dot{v} = (F \cos \alpha) / m - D / m - g \cos \theta$$

$$\dot{\theta} = (F \sin \alpha) / mv + L / mv + (g / v - v / r) \sin \theta,$$

where θ is the flight path angle from local vertical, α is the angle of attack, F is the engine thrust, D is drag, L is lift, r is the radial distance to the planet's center, and g is the acceleration due to gravity, which varies with the inverse square of the radial distance:

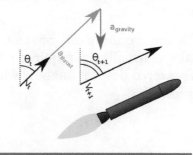

A diagram showing the velocity and force vectors acting on a space vehicle during launch.

$$g = g_0 (r_0 / r)^2$$

Mass, of course changes as propellant is consumed and rocket stages, engines or tanks are shed (if applicable). Integrating the two equations from time zero (when both v and θ are 0) gives planet-fixed values of v and θ at any time in the flight:

$$v = \int_{t_0}^{t} \dot{v}dt$$

$$\theta = \int_{t_0}^{t} \dot{\theta}dt$$

Finite element analysis can be used to numerically integrate often by breaking the flight into small time increments.

For most launch vehicles, relatively small levels of lift are generated, and a gravity turn is employed, depending mostly on the third term of the angle rate equation. But notice, when the angle is initially 0 immediately after liftoff, the only force which can cause the vehicle to pitch over is the engine thrust acting at a non-zero angle of attack (first term), until a non-zero pitch angle is attained. In the gravity turn, pitch-over is initiated by applying an increasing angle of attack (by means of gimballed engine thrust), followed by a gradual decrease in angle of attack through the remainder of the flight.

Once velocity and flight path angle are known, altitude and downrange distance are computed as:

$$h = \int_{t_0}^{t} v \cos \theta \, dt$$

$$r = r_0 + h$$

$$s = r_0 \int_{t_0}^{t} v / r \sin \theta \, dt$$

The planet-fixed values of v and θ are converted to space-fixed (inertial) values with the following conversions:

$$v_s = \sqrt{v^2 + 2\omega rv \cos \phi \sin \theta \sin A_z + (\omega r \cos \theta)^2},$$

where ω is the planet's rotational rate in radians per second, φ is the launch site latitude, and A_z is the launch azimuth angle.

$$\theta_s = \arccos(v \cos \theta / v_s)$$

Final v_s, θ_s and r must match the requirements of the target orbit as determined by orbital mechanics where final v_s is usually the required periapsis (or circular) velocity, and final θ_s is 90 degrees. A powered descent analysis would use the same procedure, with reverse boundary conditions.

Attitude Control

Attitude control is the exercise of control over the orientation of an object with respect to an inertial frame of reference or another entity (the celestial sphere, certain fields, nearby objects, etc.). The attitude of a craft can be described using three mutually perpendicular axes of rotation, generally referred to as roll, pitch, and yaw angles respectively (with the roll axis in line with the primary engine direction of thrust). Orientation can be determined by calibration using an external guidance system, such as determining the angles to a reference star or the Sun, then internally monitored using an inertial system of mechanical or optical gyroscopes. Orientation is a vector quantity described by three angles for the instantaneous direction, and the instantaneous rates of roll in all three axes of rotation. The aspect of control implies both awareness of the instantaneous orientation and rates of roll and the ability to change the roll rates to assume a new orientation using either a reaction control system or other means.

Newton's second law, applied to rotational rather than linear motion, becomes:

$$\tau = I_x \alpha,$$

where τ is the net torque (or *moment*) exerted on the vehicle, I_x is its moment of inertia about the axis of rotation, and α is the angular acceleration vector in radians per second per second. Therefore, the rotational rate in degrees per second per second is

$$\alpha = (180 / \pi)\mathbf{T} / I_x,$$

and the angular rotation rate ω (degrees per second) is obtained by integrating α over time, and the angular rotation θ is the time integral of the rate, analogous to linear motion. The three principal moments of inertia I_x, I_y, and I_z about the roll, pitch and yaw axes, are determined through the spacecraft's center of mass.

Attitude control torque, absent aerodynamic forces, is frequently applied by a reaction control system, a set of thrusters located about the vehicle. The thrusters are fired, either manually or under automatic guidance control, in short bursts to achieve the desired rate of rotation, and then fired in the opposite direction to halt rotation at the desired position. The torque about a specific axis is:

$$\tau = \sum_{i=1}^{N} (r_i \times \mathbf{F_i})$$

where $\mathbf{r}$ is its distance from the center of mass, and $\mathbf{F}$ is the thrust of an individual thruster (only the component of $\mathbf{F}$ perpendicular to $\mathbf{r}$ is included.)

For situations where propellant consumption may be a problem (such as long-duration satellites or space stations), alternative means may be used to provide the control torque, such as reaction wheels or control moment gyroscopes.

Orbital Flight

Orbital mechanics are used to calculate flight in orbit about a central body. For sufficiently high orbits (generally at least 190 kilometers (100 NM) in the case of Earth), aerodynamic force may be assumed to be negligible for relatively short term missions (though a small amount of drag may be present which results in decay of orbital energy over longer periods of time.) When the central body's mass is much larger than the spacecraft, and other bodies are sufficiently far away, the solution of orbital trajectories can be treated as a two-body problem.

This can be shown to result in the trajectory being ideally a conic section (circle, ellipse, parabola or hyperbola) with the central body located at one focus. Orbital trajectories are either circles or ellipses; the parabolic trajectory represents first escape of the vehicle from the central body's gravitational field. Hyperbolic trajectories are escape trajectories with excess velocity, and will be covered under Interplanetary flight below.

Elliptical orbits are characterized by three elements. The semi-major axis a is the average of the radius at apoapsis and periapsis:

$$a = (r_a + r_p)/2$$

The eccentricity e can then be calculated for an ellipse, knowing the apses:

$$e = r_a / a - 1$$

The time period for a complete orbit is dependent only on the semi-major axis, and is independent of eccentricity:

$$TP = 2\pi\sqrt{a^3 / \mu}$$

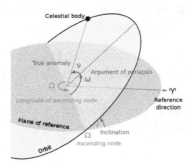

The angular orbital elements of a spacecraft orbiting a central body, defining orientation of the orbit in relation to its fundamental reference plane

The orientation of the orbit in space is specified by three angles:

- The *inclination* i, of the orbital plane with the fundamental plane (this is usually a planet or moon's equatorial plane, or in the case of a solar orbit, the Earth's orbital plane around the Sun, known as the ecliptic.) Positive inclination is northward, while negative inclination is southward.

- The *longitude of the ascending node* Ω, measured in the fundamental plane counter-clockwise looking southward, from a reference direction (usually the vernal equinox) to the line where the spacecraft crosses this plane from south to north. (If inclination is zero, this angle is undefined and taken as 0.)

- The *argument of periapsis* ω, measured in the orbital plane counter-clockwise looking southward, from the ascending node to the periapsis. If the inclination is 0, there is no ascending node, so ω is measured from the reference direction. For a circular orbit, there is no periapsis, so ω is taken as 0.

The orbital plane is ideally constant, but is usually subject to small perturbations caused by planetary oblateness and the presence of other bodies.

The spacecraft's position in orbit is specified by the *true anomaly,* ν, an angle measured from the periapsis, or for a circular orbit, from the ascending node or reference direction. The *semi-latus rectum*, or radius at 90 degrees from periapsis, is:

$$p = a(1 - e^2)$$

The radius at any position in flight is:

$$r = \frac{p}{(1 + e\cos v)}$$

and the velocity at that position is:

$$v = \sqrt{\mu(2/r - 1/a)}$$

Circular Orbit

For a circular orbit, $r_a = r_p = a$, and eccentricity is 0. Circular velocity at a given radius is:

$$v_c = \sqrt{\mu/r}$$

Elliptical Orbit

For an elliptical orbit, e is greater than 0 but less than 1. The periapsis velocity is:

$$v_p = \sqrt{\frac{\mu(1+e)}{a(1-e)}}$$

and the apoapsis velocity is:

$$v_a = \sqrt{\frac{\mu(1-e)}{a(1+e)}}$$

The limiting condition is a parabolic escape orbit, when e = 1 and r_a becomes infinite. Escape velocity at periapsis is then

$$v_e = \sqrt{2\mu/r_p}$$

Flight Path Angle

The *specific angular momentum* of any conic orbit, h, is constant, and is equal to the product of radius and velocity at periapsis. At any other point in the orbit, it is equal to:

$$h = rv\cos\phi,$$

where φ is the flight path angle measured from the local horizontal (perpendicular to r.) This allows the calculation of φ, knowing radius and velocity at any point in the orbit:

$$\phi = \arccos(\frac{r_p v_p}{rv})$$

Note that flight path angle is a constant 0 degrees (90 degrees from local vertical) for a circular orbit.

True Anomaly as a Function of Time

It can be shown that the angular momentum equation given above also relates the rate of change in true anomaly to r, v and φ, thus the true anomaly can be found as a function of time since periapsis passage by integration:

$$v = r_p v_p \int_{t_p}^{t} \frac{1}{r^2} dt$$

Conversely, the time required to reach a given anomaly is:

$$t = \frac{1}{r_p v_p} \int_{0}^{v} r^2 dv$$

Change of Plane

Sample delta-v budget will enumerate various classes of maneuvers, delta-v per maneuver, number of maneuvers required over the time of the mission.

Translunar Flight

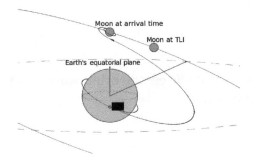

A typical translunar trajectory

Vehicles sent on lunar or planetary missions are generally not launched on a direct trajectory, but first put into a low Earth parking orbit; this allows the flexibility of a bigger launch window and more time for checking that the vehicle is in good condition for the flight. A popular misconception is that escape velocity is required for flight to the Moon; it is not. Rather, the vehicle's apogee is raised high enough to take it to a point (before it reaches apogee) where it enters the Moon's gravitational sphere of influence (though the required velocity is close to that of escape.) This is defined as the distance from a satellite at which its gravitational pull on a spacecraft equals that of its central body, which is:

$$r_{SOI} = D \left(\frac{m_s}{m_c} \right)^{2/5} ,$$

where D is the mean distance from the satellite to the central body, and m_c and m_s are the masses of the central body and satellite, respectively. This value is approximately 66,300 kilometers (35,800 NM) from Earth's Moon.

A significant portion of the vehicle's flight (other than immediate proximity to the Earth or Moon) requires accurate solution as a three-body problem, but may be preliminarily modeled as a patched conic.

Translunar Injection

This must be timed so that the Moon will be in position to capture the vehicle, and might be modeled to a first approximation as a Hohmann transfer. However, the rocket burn duration is usually long enough, and occurs during the change in flight path angle, so that this is not very accurate, requiring instead integration of a simplified version of the velocity and angle rate equations given above in Powered flight:

$$\dot{v} = (F\cos\alpha)/m - g\cos\theta$$

$$\dot{\theta} = (F\sin\alpha)/mv + (g/v - v/r)\sin\theta,$$

Mid-course Corrections

A simple lunar trajectory stays in one plane, resulting in lunar flyby or orbit within a small range of inclination to the Moon's equator. This also permits a "free return", in which the spacecraft would return to the appropriate position for reentry into the Earth's atmosphere if it were not injected into lunar orbit. Relatively small velocity changes are usually required to correct for trajectory errors. Such a trajectory was used for the Apollo 8, Apollo 10, Apollo 11, and Apollo 12 manned lunar missions.

Greater flexibility in lunar orbital or landing site coverage (at greater angles of lunar inclination) can be obtained by performing a plane change maneuver mid-flight; however, this takes away the free-return option, as the new plane would take the spacecraft's emergency return trajectory away from the Earth's atmospheric re-entry point, and leave the spacecraft in a high Earth orbit. This type of trajectory was used for the last five Apollo missions (13 through 17).

Lunar Orbit Insertion

Interplanetary Flight

In order to completely leave one planet's gravitational field to reach another, a hyperbolic trajectory relative to the departure planet is necessary, with excess velocity added to (or subtracted from) the departure planet's orbital velocity around the Sun. The desired heliocentric transfer orbit to an outer planet will have its perihelion at the departure planet, requiring the hyperbolic excess velocity to be applied in the posigrade direction, when the spacecraft is away from the Sun. To an inner destination planet, aphelion will be at the departure planet, and the excess velocity is applied in the retrograde direction when the spacecraft is toward the Sun. Since interplanetary spacecraft spend a large period of time in the heliocentric orbit between the planets, which are at relatively large distances away, the patched-conic approximation is much more accurate for interplanetary trajectories than for translunar trajectories. The patch point between the hyperbolic trajectory relative to the departure planet and the heliocentric transfer orbit can be assumed to occur at the planet's sphere of influence radius relative to the Sun, as defined above in Orbital flight.

Hyperbolic Departure

Once the required excess velocity v_∞ (sometimes called *characteristic velocity*) is determined, the injection velocity at periapsis for a hyperbola is:

$$v_p = \sqrt{2\mu / r_p + v_\infty^2}$$

The excess velocity vector for a hyperbola is displaced from the periapsis tangent by a characteristic angle, therefore the periapsis injection burn must lead the planetary departure point by the same angle:

$$\delta = \arcsin(1/e)$$

The geometric equation for eccentricity of an ellipse cannot be used for a hyperbola. But the eccentricity can be calculated from dynamics formulations as:

$$e = \sqrt{1 + \frac{2\,h^2}{\mu^2}},$$

where h is the specific angular momentum as given above in the Orbital flight section, calculated at the periapsis:

$$h = r_p v_p,$$

and ε is the specific energy:

$$= v^2/2 - \mu/r$$

Also, the equations for r and v given in Orbital flight depend on the semi-major axis, and thus are unusable for an escape trajectory. But setting radius at periapsis equal to the r equation at zero anomaly gives an alternate expression for the semi-latus rectum:

$$p = r_p(1+e),$$

which gives a more general equation for radius versus anomaly which is usable at any eccentricity:

$$r = \frac{r_p(1+e)}{(1 + e\cos v)}$$

Substituting the alternate expression for p also gives an alternate expression for a (which is defined for a hyperbola, but no longer represents the semi-major axis). This gives an equation for velocity versus radius which is likewise usable at any eccentricity:

$$v = \sqrt{\mu\left(\frac{2}{r} - \frac{(1-e^2)}{r_p(1+e)}\right)}$$

The equations for flight path angle and anomaly versus time given in Orbital flight are also usable for hyperbolic trajectories.

Launch Windows

There is a great deal of variation with time of the velocity change required for a mission, because of the constantly varying relative positions of the planets. Therefore, optimum launch windows are often chosen from the results of porkchop plots that show contours of characteristic energy (v_∞^2) plotted versus departure and arrival time.

Atmospheric Entry

Atmospheric entry is the movement of human-made or natural objects as they enter the atmosphere of a celestial body from outer space—in the case of Earth from an altitude above the Kármán Line, (100 km). This topic is heavily concerned with the process of controlled *r*eentry of vehicles which are intended to reach the planetary surface intact, but the topic also includes uncontrolled (or minimally controlled) cases, such as the intentionally or circumstantially occurring, destructive deorbiting of satellites and the falling back to the planet of "space junk" due to orbital decay.

Flight Dynamics (Fixed-Wing Aircraft)

Aerospace engineers develop control systems for a vehicle's orientation (attitude) about its center of mass. The control systems include actuators, which exert forces in various directions, and generate rotational forces or moments about the aerodynamic center of the aircraft, and thus rotate the aircraft in pitch, roll, or yaw. For example, a pitching moment is a vertical force applied at a distance forward or aft from the aerodynamic center of the aircraft, causing the aircraft to pitch up or down.

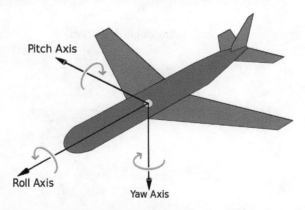

Flight dynamics is the science of air vehicle orientation and control in three dimensions. The three critical flight dynamics parameters are the angles of rotation in three dimensions about the vehicle's center of mass, known as *pitch, roll* and *yaw*.

Roll, pitch and yaw refer to rotations about the respective axes starting from a defined steady flight equilibrium state. The equilibrium roll angle is known as wings level or zero bank angle, equivalent to a level heeling angle on a ship. Yaw is known as "heading". The equilibrium pitch angle in submarine and airship parlance is known as "trim", but in aircraft, this usually refers to angle of attack, rather than orientation. However, common usage ignores this distinction between equilibrium and dynamic cases.

The most common aeronautical convention defines the roll as acting about the longitudinal axis, positive with the starboard (right) wing down. The yaw is about the vertical body axis, positive with the nose to starboard. Pitch is about an axis perpendicular to the longitudinal plane of symmetry, positive nose up.

A fixed-wing aircraft increases or decreases the lift generated by the wings when it pitches nose up or down by increasing or decreasing the angle of attack (AOA). The roll angle is also known as bank angle on a fixed-wing aircraft, which usually "banks" to change the horizontal direction of flight. An aircraft is usually streamlined from nose to tail to reduce drag making it typically advantageous to keep the sideslip angle near zero, though there are instances when an aircraft may be deliberately "sideslipped" for example a slip in a fixed-wing aircraft.

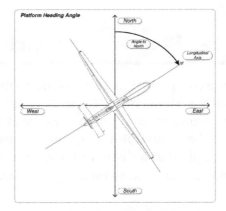

yaw or heading angle definition

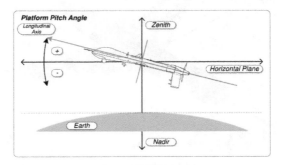

pitch angle definition

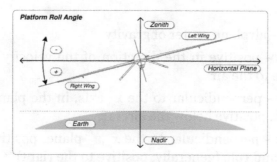

roll angle definition

Introduction

Reference Frames

Three right-handed, Cartesian coordinate systems see frequent use in flight dynamics. The first coordinate system has an origin fixed in the reference frame of the Earth:

- Earth frame

    o Origin - arbitrary, fixed relative to the surface of the Earth

    o x_E axis - positive in the direction of north

    o y_E axis - positive in the direction of east

    o z_E axis - positive towards the center of the Earth

In many flight dynamics applications, the Earth frame is assumed to be inertial with a flat x_E,y_E-plane, though the Earth frame can also be considered a spherical coordinate system with origin at the center of the Earth.

The other two reference frames are body-fixed, with origins moving along with the aircraft, typically at the center of gravity. For an aircraft that is symmetric from right-to-left, the frames can be defined as:

- Body frame

    o Origin - airplane center of gravity

    o x_b axis - positive out the nose of the aircraft in the plane of symmetry of the aircraft

    o z_b axis - perpendicular to the x_b axis, in the plane of symmetry of the aircraft, positive below the aircraft

    o y_b axis - perpendicular to the x_b,z_b-plane, positive determined by the right-hand rule (generally, positive out the right wing)

- Wind frame
  - Origin - airplane center of gravity
  - x_w axis - positive in the direction of the velocity vector of the aircraft relative to the air
  - z_w axis - perpendicular to the x_w axis, in the plane of symmetry of the aircraft, positive below the aircraft
  - y_w axis - perpendicular to the x_w,z_w-plane, positive determined by the right hand rule (generally, positive to the right)

Asymmetric aircraft have analogous body-fixed frames, but different conventions must be used to choose the precise directions of the x and z axes.

The Earth frame is a convenient frame to express aircraft translational and rotational kinematics. The Earth frame is also useful in that, under certain assumptions, it can be approximated as inertial. Additionally, one force acting on the aircraft, weight, is fixed in the $+z_E$ direction.

The body frame is often of interest because the origin and the axes remain fixed relative to the aircraft. This means that the relative orientation of the Earth and body frames describes the aircraft attitude. Also, the direction of the force of thrust is generally fixed in the body frame, though some aircraft can vary this direction, for example by thrust vectoring.

The wind frame is a convenient frame to express the aerodynamic forces and moments acting on an aircraft. In particular, the net aerodynamic force can be divided into components along the wind frame axes, with the drag force in the $-x_w$ direction and the lift force in the $-z_w$ direction.

In addition to defining the reference frames, the relative orientation of the reference frames can be determined. The relative orientation can be expressed in a variety of forms, including:

- Direction cosine or rotation matrices
- Euler angles
- Quaternions

The various Euler angles relating the three reference frames are important to flight dynamics. Many Euler angle conventions exist, but all of the rotation sequences presented below use the z-y'-x" convention. This convention corresponds to a type of Tait-Bryan angles, which are commonly referred to as Euler angles. This convention is described in detail below for the roll, pitch, and yaw Euler angles that describe the body frame orientation relative to the Earth frame. The other sets of Euler angles are described below by analogy.

To transform from the Earth frame to the body frame using Euler angles, the following rotations are done in the order prescribed. First, rotate the Earth frame axes x_E and y_E

around the z_E axis by the yaw angle ψ. This results in an intermediate reference frame with axes denoted x', y', z', where $z' = z_E$. Second, rotate the x' and z' axes around the y' axis by the pitch angle θ. This results in another intermediate reference frame with axes denoted x'', y'', z'', where $y'' = y'$. Finally, rotate the y'' and z'' axes around the x'' axis by the roll angle φ. The reference frame that results after the three rotations is the body frame.

Based on the rotations and axes conventions above, the yaw angle ψ is the angle between north and the projection of the aircraft longitudinal axis onto the horizontal plane, the pitch angle θ is the angle between the aircraft longitudinal axis and horizontal, and the roll angle φ represents a rotation around the aircraft longitudinal axis after rotating by yaw and pitch.

To transform from the Earth frame to the wind frame, the three Euler angles are the bank angle μ, the flight path angle γ, and the heading angle σ. When performing the rotations described above to obtain the wind frame from the Earth frame, (μ, γ, σ) are analogous to (φ, θ, ψ), respectively. The heading angle σ is the angle between north and the horizontal component of the velocity vector, which describes which direction the aircraft is moving relative to cardinal directions. The flight path angle γ is the angle between horizontal and the velocity vector, which describes whether the aircraft is climbing or descending. The bank angle μ represents a rotation of the lift force around the velocity vector, which may indicate whether the airplane is turning.

To transform from the wind frame to the body frame, the two Euler angles are the angle of attack α and the sideslip angle β. When performing the rotations described earlier to obtain the body frame from the wind frame, (α, β) are analogous to (θ, ψ), respectively; the angle analogous to φ in this transformation is always zero. The sideslip angle β is the angle between the velocity vector and the projection of the aircraft longitudinal axis onto the x_w, y_w-plane, which describes whether there is a lateral component to the aircraft velocity, also known as sideslip. The angle of attack α is the angle between the x_w, y_w-plane and the aircraft longitudinal axis and, among other things, is an important variable in determining the magnitude of the force of lift.

Design Cases

In analyzing the stability of an aircraft, it is usual to consider perturbations about a nominal steady flight state. So the analysis would be applied, for example, assuming:

> Straight and level flight
>
> Turn at constant speed
>
> Approach and landing
>
> Takeoff

The speed, height and trim angle of attack are different for each flight condition, in addition, the aircraft will be configured differently, e.g. at low speed flaps may be deployed and the undercarriage may be down.

Except for asymmetric designs (or symmetric designs at significant sideslip), the longitudinal equations of motion (involving pitch and lift forces) may be treated independently of the lateral motion (involving roll and yaw).

The following considers perturbations about a nominal straight and level flight path.

To keep the analysis (relatively) simple, the control surfaces are assumed fixed throughout the motion, this is stick-fixed stability. Stick-free analysis requires the further complication of taking the motion of the control surfaces into account.

Furthermore, the flight is assumed to take place in still air, and the aircraft is treated as a rigid body.

Forces of Flight

Three forces act on an aircraft in flight: weight, thrust, and the aerodynamic force.

Aerodynamic Force

Components of The Aerodynamic Force

The expression to calculate the aerodynamic force is:

$$\mathbf{F}_A = \int_{\Sigma}(-\Delta p \mathbf{n} + \mathbf{f})d\sigma$$

where:

$\Delta p \equiv$ Difference between static pressure and free current pressure

$\mathbf{n} \equiv$ outer normal vector of the element of area

$\mathbf{f} \equiv$ tangential stress vector practised by the air on the body

$\Sigma \equiv$ adequate reference surface

projected on wind axes we obtain:

$$\mathbf{F}_A = -(\mathbf{i}_w D + \mathbf{j}_w Q + \mathbf{k}_w L)$$

where:

$D \equiv$ Drag

$Q \equiv$ Lateral

$L \equiv$ Lift

Aerodynamic Coefficients

Dynamic pressure of the free current $\equiv q = \frac{1}{2}\rho V^2$

Proper reference surface (wing surface, in case of planes) $\equiv S$

Pressure coefficient $\equiv C_p = \dfrac{p - p_\infty}{q}$

Friction coefficient $\equiv C_f = \dfrac{f}{q}$

Drag coefficient $\equiv C_d = \dfrac{D}{qS} = -\dfrac{1}{S}\int_\Sigma [(-C_p)\mathbf{n}\bullet\mathbf{i_w} + C_f\mathbf{t}\bullet\mathbf{i_w}]d\sigma$

Lateral force coefficient $\equiv C_Q = \dfrac{Q}{qS} = -\dfrac{1}{S}\int_\Sigma [(-C_p)\mathbf{n}\bullet\mathbf{j_w} + C_f\mathbf{t}\bullet\mathbf{j_w}]d\sigma$

Lift coefficient $\equiv C_L = \dfrac{L}{qS} = -\dfrac{1}{S}\int_\Sigma [(-C_p)\mathbf{n}\bullet\mathbf{k_w} + C_f\mathbf{t}\bullet\mathbf{k_w}]d\sigma$

Dimensionless Parameters and Aerodynamic Regimes

In absence of thermal effects, there are three remarkable dimensionless numbers:

- Compressibility of the flow:

    Mach number $\equiv M = \dfrac{V}{a}$
- Viscosity of the flow:

    Reynolds number $\equiv Re = \dfrac{\rho Vl}{\mu}$
- Rarefaction of the flow:

    Knudsen number $\equiv Kn = \dfrac{\lambda}{l}$

where:

$a = \sqrt{kR\theta} \equiv$ speed of sound

$R \equiv$ gas constant by mass unity

$\theta \equiv$ absolute temperature

$\lambda = \dfrac{\mu}{\rho}\sqrt{\dfrac{\pi}{2R\theta}} = \dfrac{M}{Re}\sqrt{\dfrac{k\pi}{2}} \equiv$ mean free path

According to λ there are three possible rarefaction grades and their corresponding motions are called:

- Continuum current (negligible rarefaction): $\dfrac{M}{Re} \ll 1$

- Transition current (moderate rarefaction): $\dfrac{\quad}{Re} \approx$

- Free molecular current (high rarefaction): $\dfrac{M}{Re} \gg 1$

The motion of a body through a flow is considered, in flight dynamics, as continuum current. In the outer layer of the space that surrounds the body viscosity will be negligible. However viscosity effects will have to be considered when analysing the flow in the nearness of the boundary layer.

Depending on the compressibility of the flow, different kinds of currents can be considered:

- Incompressible subsonic current: $0 < M < 0.3$

- Compressible subsonic current: $0.3 < M < 0.8$

- Transonic current: $0.8 < M < 1.2$

- Supersonic current: $1.2 < M < 5$

- Hypersonic current: $5 < M$

Drag Coefficient Equation and Aerodynamic Efficiency

If the geometry of the body is fixed and in case of symmetric flight (β=0 and Q=0), pressure and friction coefficients are functions depending on:

$$C_p = C_p(\alpha, M, Re, P)$$
$$C_f = C_f(\alpha, M, Re, P)$$

where:

$\alpha \equiv$ angle of attack

$P \equiv$ considered point of the surface

Under these conditions, drag and lift coefficient are functions depending exclusively on the angle of attack of the body and Mach and Reynolds numbers. Aerodynamic efficiency, defined as the relation between lift and drag coefficients, will depend on those parameters as well.

$$\begin{cases} C_D = C_D(\alpha, M, Re) \\ C_L = C_L(\alpha, M, Re) \\ E = E(\alpha, M, Re) = \dfrac{C_L}{C_D} \end{cases}$$

It is also possible to get the dependency of the drag coefficient respect to the lift coefficient. This relation is known as the drag coefficient equation:

$$C_D = C_D(C_L, M, Re) \equiv \text{drag coefficient equation}$$

The aerodynamic efficiency has a maximum value, E_{max}, respect to C_L where the tangent line from the coordinate origin touches the drag coefficient equation plot.

The drag coefficient, C_D, can be decomposed in two ways. First typical decomposition separates pressure and friction effects:

$$C_D = C_{Df} + C_{Dp} \begin{cases} C_{Df} = \dfrac{D}{qS} = -\dfrac{1}{S}\int_\Sigma C_f \mathbf{t} \cdot \mathbf{i}_w \, d\sigma \\[2mm] C_{Dp} = \dfrac{D}{qS} = -\dfrac{1}{S}\int_\Sigma (-C_p) \mathbf{n} \cdot \mathbf{i}_w \, d\sigma \end{cases}$$

There's a second typical decomposition taking into account the definition of the drag coefficient equation. This decomposition separates the effect of the lift coefficient in the equation, obtaining two terms C_{Do} and C_{Di}. C_{Do} is known as the parasitic drag coefficient and it is the base draft coefficient at zero lift. C_{Di} is known as the induced drag coefficient and it is produced by the body lift.

$$C_D = C_{D0} + C_{Di} \begin{cases} C_{D0} = (C_D)_{C_L=0} \\[2mm] C_{Di} \end{cases}$$

Parabolic and Generic Drag Coefficient

A good attempt for the induced drag coefficient is to assume a parabolic dependency of the lift

$$C_{Di} = kC_L^2 \Rightarrow C_D = C_{D0} + kC_L^2$$

Aerodynamic efficiency is now calculated as:

$$E = \frac{C_L}{C_{D0} + kC_L^2} \Rightarrow \begin{cases} E_{max} = \dfrac{1}{2\sqrt{kC_{D0}}} \\[3mm] (C_L)_{Emax} = \sqrt{\dfrac{C_{D0}}{k}} \\[3mm] (C_{Di})_{Emax} = C_{D0} \end{cases}$$

If the configuration of the pane is symmetrical respect to the XY plane, minimum drag coefficient equals to the parasitic drag of the plane.

$$C_{Dmin} = (C_D)_{CL=0} = C_{D0}$$

In case the configuration is asymmetrical respect to the XY plane, however, minimum flag differs from the parasitic drag. On these cases, a new approximate parabolic drag equation can be traced leaving the minimum drag value at zero lift value.

$$C_{Dmin} = C_{DM} \neq (C_D)_{CL=0}$$
$$C_D = C_{DM} + k(C_L - C_{LM})^2$$

Variation of Parameters with the Mach Number

The Coefficient of pressure varies with Mach number by the relation given below:

$$C_p = \frac{C_{p0}}{\sqrt{|1 - M_\infty^2|}}$$

where

- C_p is the compressible pressure coefficient
- C_{po} is the incompressible pressure coefficient
- M_∞ is the freestream Mach number.

This relation is reasonably accurate for 0.3 < M < 0.7 and when $M = 1$ it becomes ∞ which is impossible physical situation and is called Prandtl–Glauert singularity.

Directional Stability

Directional or weathercock stability is concerned with the static stability of the airplane about the z axis. Just as in the case of longitudinal stability it is desirable that the aircraft should tend to return to an equilibrium condition when subjected to some form of yawing disturbance. For this the slope of the yawing moment curve must be positive. An airplane possessing this mode of stability will always point towards the relative wind, hence the name weathercock stability.

Dynamic Stability and Control

Longitudinal Modes

It is common practice to derive a fourth order characteristic equation to describe the longitudinal motion, and then factorize it approximately into a high frequency mode and a low frequency mode. The approach adopted here is using qualitative knowledge of aircraft behavior to simplify the equations from the outset, reaching the result by a more accessible route.

The two longitudinal motions (modes) are called the short period pitch oscillation (SPPO), and the phugoid.

Short-period Pitch Oscillation

A short input (in control systems terminology an impulse) in pitch (generally via the elevator in a standard configuration fixed-wing aircraft) will generally lead to overshoots about the trimmed condition. The transition is characterized by a damped simple harmonic motion about the new trim. There is very little change in the trajectory over the time it takes for the oscillation to damp out.

Generally this oscillation is high frequency (hence short period) and is damped over a period of a few seconds. A real-world example would involve a pilot selecting a new climb attitude, for example 5° nose up from the original attitude. A short, sharp pull back on the control column may be used, and will generally lead to oscillations about the new trim condition. If the oscillations are poorly damped the aircraft will take a long period of time to settle at the new condition, potentially leading to Pilot-induced oscillation. If the short period mode is unstable it will generally be impossible for the pilot to safely control the aircraft for any period of time.

This damped harmonic motion is called the short period pitch oscillation, it arises from the tendency of a stable aircraft to point in the general direction of flight. It is very similar in nature to the weathercock mode of missile or rocket configurations. The motion involves mainly the pitch attitude θ (theta) and incidence α (alpha). The direction of the velocity vector, relative to inertial axes is $\theta - \alpha$.. The velocity vector is:

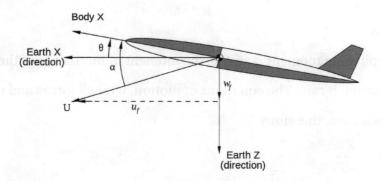

Longitudinal Equations of Motion

$$u_f = U \cos(\theta - \alpha)$$

$$w_f = U \sin(\theta - \alpha)$$

where u_f, w_f are the inertial axes components of velocity. According to Newton's Second Law, the accelerations are proportional to the forces, so the forces in inertial axes are:

$$X_f = m\frac{du_f}{dt} = m\frac{dU}{dt}\cos(\theta - \alpha) - mU\frac{d(\theta - \alpha)}{dt}\sin(\theta - \alpha)$$

$$Z = m\frac{dw}{dt} = m\frac{dU}{dt}\sin(\theta - \alpha) + mU\frac{d(\theta\ \alpha)}{dt}\cos(\theta - \alpha)$$

where m is the mass. By the nature of the motion, the speed variation $m\dfrac{dU}{dt}$ is negligible over the period of the oscillation, so:

$$X_f = -mU\frac{d(\theta - \alpha)}{dt}\sin(\theta - \alpha)$$

$$Z_f = mU\frac{d(\theta - \alpha)}{dt}\cos(\theta - \alpha)$$

But the forces are generated by the pressure distribution on the body, and are referred to the velocity vector. But the velocity (wind) axes set is not an inertial frame so we must resolve the fixed axes forces into wind axes. Also, we are only concerned with the force along the z-axis:

$$Z = -Z_f \cos(\theta - \alpha) + X_f \sin(\theta - \alpha)$$

Or:

$$Z = -mU\frac{d(\theta - \alpha)}{dt}$$

In words, the wind axes force is equal to the centripetal acceleration.

The moment equation is the time derivative of the angular momentum:

$$M = B\frac{d^2\theta}{dt^2}$$

where M is the pitching moment, and B is the moment of inertia about the pitch axis.

Let: $\dfrac{\ }{dt}$, the pitch rate. The equations of motion, with all forces and moments re ferred to wind axes are, therefore:

$$\frac{d\alpha}{dt} = q + \frac{Z}{mU}$$

$$\frac{dq}{dt} = \frac{M}{B}$$

We are only concerned with perturbations in forces and moments, due to perturbations in the states α and q, and their time derivatives. These are characterized by stability derivatives determined from the flight condition. The possible stability derivatives are:

Z_α Lift due to incidence, this is negative because the z-axis is downwards whilst positive incidence causes an upwards force.

Z_q Lift due to pitch rate, arises from the increase in tail incidence, hence is also negative, but small compared with Z_α.

M_α Pitching moment due to incidence - the static stability term. Static stability requires this to be negative.

M_q Pitching moment due to pitch rate - the pitch damping term, this is always negative.

Since the tail is operating in the flowfield of the wing, changes in the wing incidence cause changes in the downwash, but there is a delay for the change in wing flowfield to affect the tail lift, this is represented as a moment proportional to the rate of change of incidence:

$$M_{\dot\alpha}$$

Increasing the wing incidence without increasing the tail incidence produces a nose up moment, so $M_{\dot\alpha}$ is expected to be positive.

The equations of motion, with small perturbation forces and moments become:

$$\frac{d\alpha}{dt} = \left(1 + \frac{Z_q}{mU}\right)q + \frac{Z_\alpha}{mU}\alpha$$

$$\frac{dq}{dt} = \frac{M_q}{B}q + \frac{M_\alpha}{B}\alpha + \frac{M_{\dot\alpha}}{B}\dot\alpha$$

These may be manipulated to yield as second order linear differential equation in α :

$$\frac{d^2\alpha}{dt^2} - \left(\frac{Z_\alpha}{mU} + \frac{M_q}{B} + (1 + \frac{Z_q}{mU})\frac{M_{\dot\alpha}}{B}\right)\frac{d\alpha}{dt} + \left(\frac{Z_\alpha}{mU}\frac{M_q}{B} - \frac{M_\alpha}{B}(1 + \frac{Z_q}{mU})\right)\alpha = 0$$

This represents a damped simple harmonic motion.

We should expect $\dfrac{Z_q}{mU}$ to be small compared with unity, so the coefficient of α (the 'stiffness' term) will be positive, provided $M_\alpha < \dfrac{Z_\alpha}{mU}M_q$. This expression is dominated by M_α, which defines the longitudinal static stability of the aircraft, it must be negative for stability. The damping term is reduced by the downwash effect, and it is difficult to design an aircraft with both rapid natural response and heavy damping. Usually, the response is underdamped but stable.

Phugoid

If the stick is held fixed, the aircraft will not maintain straight and level flight, but will start to dive, level out and climb again. It will repeat this cycle until the pilot intervenes. This long period oscillation in speed and height is called the phugoid mode. This is analyzed by assuming that the SSPO performs its proper function and maintains the angle of attack near its nominal value. The two states which are mainly affected are the climb angle γ (gamma) and speed. The small perturbation equations of motion are:

$$mU\frac{d\gamma}{dt} = -Z$$

which means the centripetal force is equal to the perturbation in lift force.

For the speed, resolving along the trajectory:

$$m\frac{du}{dt} = X - mg\gamma$$

where g is the acceleration due to gravity at the earths surface. The acceleration along the trajectory is equal to the net x-wise force minus the component of weight. We should not expect significant aerodynamic derivatives to depend on the climb angle, so only X_u and Z_u need be considered. X_u is the drag increment with increased speed, it is negative, likewise Z_u is the lift increment due to speed increment, it is also negative because lift acts in the opposite sense to the z-axis.

The equations of motion become:

$$mU\frac{d\gamma}{dt} = -Z_u u$$

$$m\frac{du}{dt} = X_u u - mg\gamma$$

These may be expressed as a second order equation in climb angle or speed perturbation:

$$\frac{d^2u}{dt^2} - \frac{X_u}{m}\frac{du}{dt} - \frac{Z_u g}{mU}u = 0$$

Now lift is very nearly equal to weight:

$$Z = \frac{1}{2}\rho U^2 c_L S_w = W$$

where ρ is the air density, S_w is the wing area, W the weight and c_L is the lift coefficient (assumed constant because the incidence is constant), we have, approximately:

$$Z_u = \frac{2W}{U} = \frac{2mg}{U}$$

The period of the phugoid, T, is obtained from the coefficient of u:

$$\frac{2\pi}{T} = \sqrt{\frac{2g^2}{U^2}}$$

Or:

$$T = \frac{2\pi U}{\sqrt{2g}}$$

Since the lift is very much greater than the drag, the phugoid is at best lightly damped. A propeller with fixed speed would help. Heavy damping of the pitch rotation or a large rotational inertia increase the coupling between short period and phugoid modes, so that these will modify the phugoid.

Lateral Modes

With a symmetrical rocket or missile, the directional stability in yaw is the same as the pitch stability; it resembles the short period pitch oscillation, with yaw plane equivalents to the pitch plane stability derivatives. For this reason pitch and yaw directional stability are collectively known as the "weathercock" stability of the missile.

Aircraft lack the symmetry between pitch and yaw, so that directional stability in yaw is derived from a different set of stability derivatives. The yaw plane equivalent to the short period pitch oscillation, which describes yaw plane directional stability is called Dutch roll. Unlike pitch plane motions, the lateral modes involve both roll and yaw motion.

Dutch Roll

It is customary to derive the equations of motion by formal manipulation in what, to the engineer, amounts to a piece of mathematical sleight of hand. The current approach follows the pitch plane analysis in formulating the equations in terms of concepts which are reasonably familiar.

Applying an impulse via the rudder pedals should induce Dutch roll, which is the oscillation in roll and yaw, with the roll motion lagging yaw by a quarter cycle, so that the wing tips follow elliptical paths with respect to the aircraft.

The yaw plane translational equation, as in the pitch plane, equates the centripetal acceleration to the side force.

$$\frac{d\beta}{dt} = \frac{Y}{mU} - r$$

where β (beta) is the sideslip angle, Y the side force and r the yaw rate.

The moment equations are a bit trickier. The trim condition is with the aircraft at an angle of attack with respect to the airflow. The body x-axis does not align with the velocity vector, which is the reference direction for wind axes. In other words, wind axes are not principal axes (the mass is not distributed symmetrically about the yaw and roll axes). Consider the motion of an element of mass in position -z, x in the direction of the y-axis, i.e. into the plane of the paper.

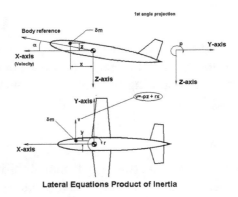

Lateral Equations Product of Inertia

If the roll rate is p, the velocity of the particle is:

$$v = -pz + xr$$

Made up of two terms, the force on this particle is first the proportional to rate of v change, the second is due to the change in direction of this component of velocity as the body moves. The latter terms gives rise to cross products of small quantities (pq, pr,qr), which are later discarded. In this analysis, they are discarded from the outset for the sake of clarity. In effect, we assume that the direction of the velocity of the particle due to the simultaneous roll and yaw rates does not change significantly throughout the motion. With this simplifying assumption, the acceleration of the particle becomes:

$$\frac{dv}{dt} = -\frac{dp}{dt}z + \frac{dr}{dt}x$$

The yawing moment is given by:

$$\delta m x \frac{dv}{dt} = -\frac{dp}{dt}xz\delta m + \frac{dr}{dt}x^2\delta m$$

There is an additional yawing moment due to the offset of the particle in the y direction:

The yawing moment is found by summing over all particles of the body:

$$N = -\frac{dp}{dt}\int xz \, dm + \frac{dr}{dt}\int x^2 + y^2 \, dm = -E\frac{dp}{dt} + C\frac{dr}{dt}$$

where N is the yawing moment, E is a product of inertia, and C is the moment of inertia about the yaw axis. A similar reasoning yields the roll equation:

$$L = A\frac{dp}{dt} - E\frac{dr}{dt}$$

where L is the rolling moment and A the roll moment of inertia.

Lateral and Longitudinal Stability Derivatives

The states are β (sideslip), r (yaw rate) and p (roll rate), with moments N (yaw) and L (roll), and force Y (sideways). There are nine stability derivatives relevant to this motion, the following explains how they originate. However a better intuitive understanding is to be gained by simply playing with a model airplane, and considering how the forces on each component are affected by changes in sideslip and angular velocity:

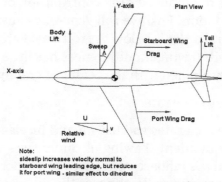

Additional Contributions to Lateral Stability

Y_β Side force due to side slip (in absence of yaw).

Sideslip generates a sideforce from the fin and the fuselage. In addition, if the wing has dihedral, side slip at a positive roll angle increases incidence on the starboard wing and reduces it on the port side, resulting in a net force component directly opposite to the sideslip direction. Sweep back of the wings has the same effect on incidence, but since the wings are not inclined in the vertical plane, backsweep alone does not affect Y_β. However, anhedral may be used with high backsweep angles in high performance aircraft to offset the wing incidence effects of sideslip. Oddly enough this does not reverse the sign of the wing configuration's contribution to Y_β (compared to the dihedral case).

Y_p Side force due to roll rate.

Roll rate causes incidence at the fin, which generates a corresponding side force. Also, positive roll (starboard wing down) increases the lift on the starboard wing and reduces it on the port. If the wing has dihedral, this will result in a side force momentarily opposing the resultant sideslip tendency. Anhedral wing and or stabilizer configurations can cause the sign of the side force to invert if the fin effect is swamped.

Y_r Side force due to yaw rate.

Yawing generates side forces due to incidence at the rudder, fin and fuselage.

N_β Yawing moment due to sideslip forces.

Sideslip in the absence of rudder input causes incidence on the fuselage and empennage, thus creating a yawing moment counteracted only by the directional stiffness which would tend to point the aircraft's nose back into the wind in horizontal flight conditions. Under sideslip conditions at a given roll angle N_β will tend to point the nose into the sideslip direction even without rudder input, causing a downward spiraling flight.

N_p Yawing moment due to roll rate.

Roll rate generates fin lift causing a yawing moment and also differentially alters the lift on the wings, thus affecting the induced drag contribution of each wing, causing a (small) yawing moment contribution. Positive roll generally causes positive N_p values unless the empennage is anhedral or fin is below the roll axis. Lateral force components resulting from dihedral or anhedral wing lift differences has little effect on N_p because the wing axis is normally closely aligned with the center of gravity.

N_r Yawing moment due to yaw rate.

Yaw rate input at any roll angle generates rudder, fin and fuselage force vectors which dominate the resultant yawing moment. Yawing also increases the speed of the outboard wing whilst slowing down the inboard wing, with corresponding changes in drag causing a (small) opposing yaw moment. N_r opposes the inherent directional stiffness which tends to point the aircraft's nose back into the wind and always matches the sign of the yaw rate input.

L_β Rolling moment due to sideslip.

A positive sideslip angle generates empennage incidence which can cause positive or negative roll moment depending on its configuration. For any non-zero sideslip angle dihedral wings causes a rolling moment which tends to return the aircraft to the horizontal, as does back swept wings. With highly swept wings the resultant rolling moment may be excessive for all stability requirements and anhedral could be used to offset the effect of wing sweep induced rolling moment.

L_r Rolling moment due to yaw rate.

Yaw increases the speed of the outboard wing whilst reducing speed of the inboard one, causing a rolling moment to the inboard side. The contribution of the fin normally supports this inward rolling effect unless offset by anhedral stabilizer above the roll axis (or dihedral below the roll axis).

L_p Rolling moment due to roll rate.

Roll creates counter rotational forces on both starboard and port wings whilst also generating such forces at the empennage. These opposing rolling moment effects have to be overcome by the aileron input in order to sustain the roll rate. If the roll is stopped at a non-zero roll angle the L_β *upward* rolling moment induced by the ensuing sideslip should return the aircraft to the horizontal unless exceeded in turn by the *downward* L_r rolling moment resulting from sideslip induced yaw rate. Longitudinal stability could be ensured or improved by minimizing the latter effect.

Equations of Motion

Since Dutch roll is a handling mode, analogous to the short period pitch oscillation, any effect it might have on the trajectory may be ignored. The body rate r is made up of the rate of change of sideslip angle and the rate of turn. Taking the latter as zero, assuming no effect on the trajectory, for the limited purpose of studying the Dutch roll:

$$\frac{d\beta}{dt} = -r$$

The yaw and roll equations, with the stability derivatives become:

$$C\frac{dr}{dt} - E\frac{dp}{dt} = N_\beta\beta - N_r\frac{d\beta}{dt} + N_p p \text{ (yaw)}$$

$$A\frac{dp}{dt} - E\frac{dr}{dt} = L_\beta\beta - L_r\frac{d\beta}{dt} + L_p p \text{ (roll)}$$

The inertial moment due to the roll acceleration is considered small compared with the aerodynamic terms, so the equations become:

$$-C\frac{d^2\beta}{dt^2} = N_\beta\beta - N_r\frac{d\beta}{dt} + N_p p$$

$$E\frac{d^2\beta}{dt^2} = L_\beta\beta - L_r\frac{d\beta}{dt} + L_p p$$

This becomes a second order equation governing either roll rate or sideslip:

$$\left(\frac{N_p}{C}\frac{E}{A} - \frac{L_p}{A}\right)\frac{d^2\beta}{dt^2} + \left(\frac{L_p}{A}\frac{N_r}{C} - \frac{N_p}{C}\frac{L_r}{A}\right)\frac{d\beta}{dt} - \left(\frac{L_p}{A}\frac{N_\beta}{C} - \frac{L_\beta}{A}\frac{N_p}{C}\right)\beta = 0$$

The equation for roll rate is identical. But the roll angle, ϕ (phi) is given by:

$$\frac{d\phi}{dt} = p$$

If p is a damped simple harmonic motion, so is ϕ, but the roll must be in quadrature with the roll rate, and hence also with the sideslip. The motion consists of oscillations in roll and yaw, with the roll motion lagging 90 degrees behind the yaw. The wing tips trace out elliptical paths.

Stability requires the "stiffness" and "damping" terms to be positive. These are:

$$\frac{\dfrac{L_p}{A}\dfrac{N_r}{C} - \dfrac{N_p}{C}\dfrac{L_r}{A}}{\dfrac{N_p}{C}\dfrac{E}{A} - \dfrac{L_p}{A}} \text{ (damping)}$$

$$\frac{\dfrac{L_\beta}{A}\dfrac{N_p}{C} - \dfrac{L_p}{A}\dfrac{N_\beta}{C}}{\dfrac{N_p}{C}\dfrac{E}{A} - \dfrac{L_p}{A}} \text{ (stiffness)}$$

The denominator is dominated by L_p, the roll damping derivative, which is always negative, so the denominators of these two expressions will be positive.

Considering the "stiffness" term: $-L_p N_\beta$ will be positive because L_p is always negative and is positive by design. L_β is usually negative, whilst N_p is positive. Excessive dihedral can destabilize the Dutch roll, so configurations with highly swept wings require anhedral to offset the wing sweep contribution to   .

The damping term is dominated by the product of the roll damping and the yaw damping derivatives, these are both negative, so their product is positive. The Dutch roll should therefore be damped.

The motion is accompanied by slight lateral motion of the center of gravity and a more "exact" analysis will introduce terms in Y_β etc. In view of the accuracy with which stability derivatives can be calculated, this is an unnecessary pedantry, which serves to obscure the relationship between aircraft geometry and handling, which is the fundamental objective of this article.

Roll Subsidence

Jerking the stick sideways and returning it to center causes a net change in roll orientation.

The roll motion is characterized by an absence of natural stability, there are no stability derivatives which generate moments in response to the inertial roll angle. A roll disturbance induces a roll rate which is only canceled by pilot or autopilot intervention. This takes place with insignificant changes in sideslip or yaw rate, so the equation of motion reduces to:

$$A\frac{dp}{dt} = L_p p.$$

L_p is negative, so the roll rate will decay with time. The roll rate reduces to zero, but there is no direct control over the roll angle.

Spiral Mode

Simply holding the stick still, when starting with the wings near level, an aircraft will usually have a tendency to gradually veer off to one side of the straight flightpath. This is the (slightly unstable) spiral mode.

Spiral Mode Trajectory

In studying the trajectory, it is the direction of the velocity vector, rather than that of the body, which is of interest. The direction of the velocity vector when projected on to the horizontal will be called the track, denoted μ (mu). The body orientation is called the heading, denoted ψ (psi). The force equation of motion includes a component of weight:

$$\frac{d\mu}{dt} = \frac{Y}{mU} + \frac{g}{U}\phi$$

where g is the gravitational acceleration, and U is the speed.

Including the stability derivatives:

$$\frac{d\mu}{dt} = \frac{Y_\beta}{mU}\beta + \frac{Y_r}{mU}r + \frac{Y_p}{mU}p + \frac{g}{U}\phi$$

Roll rates and yaw rates are expected to be small, so the contributions of Y_r and Y_p will be ignored.

The sideslip and roll rate vary gradually, so their time derivatives are ignored. The yaw and roll equations reduce to:

$$N_\beta\beta + N_r\frac{d\mu}{dt} + N_p p = 0 \text{ (yaw)}$$

$$L_\beta\beta + L_r\frac{d\mu}{dt} + L_p p = 0 \text{ (roll)}$$

Solving for β and p:

$$\beta = \frac{(L_r N_p - L_p N_r)}{(L_p N_\beta - N_p L_\beta)}\frac{d\mu}{dt}$$

$$p = \frac{(L_\beta N_r - L_r N_\beta)}{(L_p N_\beta - N_p L_\beta)}\frac{d\mu}{dt}$$

Substituting for sideslip and roll rate in the force equation results in a first order equation in roll angle:

$$\frac{d\phi}{dt} = mg\frac{(L_\beta N_r - N_\beta L_r)}{mU(L_p N_\beta - N_p L_\beta) - Y_\beta(L_r N_p - L_p N_r)}\phi$$

This is an exponential growth or decay, depending on whether the coefficient of is positive or negative. The denominator is usually negative, which requires (both products are positive). This is in direct conflict with the Dutch roll stability requirement, and it is difficult to design an aircraft for which both the Dutch roll and spiral mode are inherently stable.

Since the spiral mode has a long time constant, the pilot can intervene to effectively stabilize it, but an aircraft with an unstable Dutch roll would be difficult to fly. It is usual to design the aircraft with a stable Dutch roll mode, but slightly unstable spiral mode.

Technology Related to Aerospace Engineering

The electronic systems that are used on aircrafts and spacecrafts are known as avionics. Devices that are used for controlling aircrafts, missiles and satellites are known as guidance systems. A guidance system also helps in the navigation of these objects. The aspects elucidated in this section are of vital importance, and it provides a better understanding of aerospace engineering.

Avionics

Avionics are the electronic systems used on aircraft, artificial satellites, and spacecraft. Avionic systems include communications, navigation, the display and management of multiple systems, and the hundreds of systems that are fitted to aircraft to perform individual functions. These can be as simple as a searchlight for a police helicopter or as complicated as the tactical system for an airborne early warning platform. The term *avionics* is a portmanteau of the words *aviation* and *electronics*.

F105 Thunderchief with avionics laid out

Radar and other avionics in the nose of a Cessna Citation I/SP

History

The term avionics was coined by the journalist Philip J. Klass as a portmanteau of aviation electronics. Many modern avionics have their origins in World War II wartime developments. For example, autopilot systems that are prolific today were started to help bomber planes fly steadily enough to hit precision targets from high altitudes. Famously, radar was developed in the UK, Germany, and the United States during the same period. Modern avionics is a substantial portion of military aircraft spending. Aircraft like the F15E and the now retired F14 have roughly 20 percent of their budget spent on avionics. Most modern helicopters now have budget splits of 60/40 in favour of avionics.

The civilian market has also seen a growth in cost of avionics. Flight control systems (fly-by-wire) and new navigation needs brought on by tighter airspaces, have pushed up development costs. The major change has been the recent boom in consumer flying. As more people begin to use planes as their primary method of transportation, more elaborate methods of controlling aircraft safely in these high restrictive airspaces have been invented.

Modern Avionics

Avionics plays a heavy role in modernization initiatives like the Federal Aviation Administration's (FAA) Next Generation Air Transportation System project in the United States and the Single European Sky ATM Research (SESAR) initiative in Europe. The Joint Planning and Development Office put forth a roadmap for avionics in six areas:

- Published Routes and Procedures – Improved navigation and routing

- Negotiated Trajectories – Adding data communications to create preferred routes dynamically

- Delegated Separation – Enhanced situational awareness in the air and on the ground

- LowVisibility/CeilingApproach/Departure – Allowing operations with weather constraints with less ground infrastructure

- Surface Operations – To increase safety in approach and departure

- ATM Efficiencies – Improving the ATM process

Aircraft Avionics

The cockpit of an aircraft is a typical location for avionic equipment, including control, monitoring, communication, navigation, weather, and anti-collision systems. The majority of aircraft power their avionics using 14- or 28volt DC electrical systems; however, larger, more sophisticated aircraft (such as airliners or military combat aircraft) have AC systems operating at 400 Hz, 115 volts AC. There are several major vendors of flight avionics, including Panasonic Avionics Corporation, Honeywell (which now owns Bendix/King), Rockwell Collins, Thales Group, GE Aviation Systems, Garmin, Parker Hannifin, UTC Aerospace Systems and Avidyne Corporation.

One source of international standards for avionics equipment are prepared by the Airlines Electronic Engineering Committee (AEEC) and published by ARINC.

Communications

Communications connect the flight deck to the ground and the flight deck to the passengers. Onboard communications are provided by public-address systems and aircraft intercoms.

The VHF aviation communication system works on the airband of 118.000 MHz to 136.975 MHz. Each channel is spaced from the adjacent ones by 8.33 kHz in Europe, 25 kHz elsewhere. VHF is also used for line of sight communication such as aircraft-to-aircraft and aircraft-to-ATC. Amplitude modulation (AM) is used, and the conversation is performed in simplex mode. Aircraft communication can also take place using HF (especially for trans-oceanic flights) or satellite communication.

Navigation

Navigation is the determination of position and direction on or above the surface of the Earth. Avionics can use satellite-based systems (such as GPS and WAAS), ground-based systems (such as VOR or LORAN), or any combination thereof. Navigation systems calculate the position automatically and display it to the flight crew on moving map displays. Older avionics required a pilot or navigator to plot the intersection of signals on a paper map to determine an aircraft's location; modern systems calculate the position automatically and display it to the flight crew on moving map displays.

Monitoring

The first hints of glass cockpits emerged in the 1970s when flight-worthy cathode ray tubes (CRT) screens began to replace electromechanical displays, gauges and instruments. A "glass" cockpit refers to the use of computer monitors instead of gauges and other analog displays. Aircraft were getting progressively more displays, dials and information dashboards that eventually competed for space and pilot attention. In the 1970s, the average aircraft had more than 100 cockpit instruments and controls.

The Airbus A380 glass cockpit featuring pull-out keyboards and two wide computer screens on the sides for pilots.

Glass cockpits started to come into being with the Gulfstream GIV private jet in 1985. One of the key challenges in glass cockpits is to balance how much control is automated and how much the pilot should do manually. Generally they try to automate flight operations while keeping the pilot constantly informed.

Aircraft Flight-control Systems

Aircraft have means of automatically controlling flight. Autopilot was first invented by Lawrence Sperry during World War I to fly bomber planes steady enough to hit precision targets from 25,000 feet. When it was first adopted by the U.S. military, a Honeywell engineer sat in the back seat with bolt cutters to disconnect the autopilot in case of emergency. Nowadays most commercial planes are equipped with aircraft flight control systems in order to reduce pilot error and workload at landing or take-off.

The first simple commercial auto-pilots were used to control heading and altitude and had limited authority on things like thrust and flight control surfaces. In helicopters, auto-stabilization was used in a similar way. The first systems were electromechanical. The advent of fly by wire and electro-actuated flight surfaces (rather than the traditional hydraulic) has increased safety. As with displays and instruments, critical devices that were electro-mechanical had a finite life. With safety critical systems, the software is very strictly tested.

Collision-avoidance Systems

To supplement air traffic control, most large transport aircraft and many smaller ones use a traffic alert and collision avoidance system (TCAS), which can detect the location of nearby aircraft, and provide instructions for avoiding a midair collision. Smaller aircraft may use simpler traffic alerting systems such as TPAS, which are passive (they do not actively interrogate the transponders of other aircraft) and do not provide advisories for conflict resolution.

To help avoid controlled flight into terrain (CFIT), aircraft use systems such as ground-proximity warning systems (GPWS), which use radar altimeters as a key element. One of the major weaknesses of GPWS is the lack of "look-ahead" information, because it only provides altitude above terrain "look-down". In order to overcome this weakness, modern aircraft use a terrain awareness warning system (TAWS).

Black Boxes

Commercial aircraft cockpit data recorders, commonly known as a "black box", store flight information and audio from the cockpit. They are often recovered from a plane after a crash to determine control settings and other parameters during the incident.

Weather Systems

Weather systems such as weather radar (typically Arinc 708 on commercial aircraft) and lightning detectors are important for aircraft flying at night or in instrument meteorological conditions, where it is not possible for pilots to see the weather ahead. Heavy precipitation (as sensed by radar) or severe turbulence (as sensed by lightning activity) are both indications of strong convective activity and severe turbulence, and weather systems allow pilots to deviate around these areas.

Lightning detectors like the Stormscope or Strikefinder have become inexpensive enough that they are practical for light aircraft. In addition to radar and lightning detection, observations and extended radar pictures (such as NEXRAD) are now available through satellite data connections, allowing pilots to see weather conditions far beyond the range of their own in-flight systems. Modern displays allow weather information to be integrated with moving maps, terrain, and traffic onto a single screen, greatly simplifying navigation.

Modern weather systems also include wind shear and turbulence detection and terrain and traffic warning systems. Inplane weather avionics are especially popular in Africa, India, and other countries where air-travel is a growing market, but ground support is not as well developed.

Aircraft Management Systems

There has been a progression towards centralized control of the multiple complex systems fitted to aircraft, including engine monitoring and management. Health and usage monitoring systems (HUMS) are integrated with aircraft management computers to give maintainers early warnings of parts that will need replacement.

The integrated modular avionics concept proposes an integrated architecture with application software portable across an assembly of common hardware modules. It has been used in fourth generation jet fighters and the latest generation of airliners.

Mission or Tactical Avionics

Military aircraft have been designed either to deliver a weapon or to be the eyes and ears of other weapon systems. The vast array of sensors available to the military is used for whatever tactical means required. As with aircraft management, the bigger sensor platforms (like the E3D, JSTARS, ASTOR, Nimrod MRA4, Merlin HM Mk 1) have mission-management computers.

Police and EMS aircraft also carry sophisticated tactical sensors.

Military Communications

While aircraft communications provide the backbone for safe flight, the tactical systems are designed to withstand the rigors of the battle field. UHF, VHF Tactical (30–88 MHz) and SatCom systems combined with ECCM methods, and cryptography secure the communications. Data links such as Link 11, 16, 22 and BOWMAN, JTRS and even TETRA provide the means of transmitting data (such as images, targeting information etc.).

Radar

Airborne radar was one of the first tactical sensors. The benefit of altitude providing range has meant a significant focus on airborne radar technologies. Radars include airborne early warning (AEW), anti-submarine warfare (ASW), and even weather radar (Arinc 708) and ground tracking/proximity radar.

The military uses radar in fast jets to help pilots fly at low levels. While the civil market has had weather radar for a while, there are strict rules about using it to navigate the aircraft.

Sonar

Dipping sonar fitted to a range of military helicopters allows the helicopter to protect shipping assets from submarines or surface threats. Maritime support aircraft can drop active and passive sonar devices (sonobuoys) and these are also used to determine the location of hostile submarines.

Electro-Optics

Electro-optic systems include devices such as the head-up display (HUD), forward looking infrared (FLIR), infra-red search and track and other passive infrared devices (Passive infrared sensor). These are all used to provide imagery and information to the flight crew. This imagery is used for everything from search and rescue to navigational aids and target acquisition.

ESM/DAS

Electronic support measures and defensive aids are used extensively to gather information about threats or possible threats. They can be used to launch devices (in some cases automatically) to counter direct threats against the aircraft. They are also used to determine the state of a threat and identify it.

Aircraft Networks

The avionics systems in military, commercial and advanced models of civilian aircraft are interconnected using an avionics databus. Common avionics databus protocols, with their primary application, include:

- Aircraft Data Network (ADN): Ethernet derivative for Commercial Aircraft
- Avionics Full-Duplex Switched Ethernet (AFDX): Specific implementation of ARINC 664 (ADN) for Commercial Aircraft
- ARINC 429: Generic Medium-Speed Data Sharing for Private and Commercial Aircraft
- ARINC 664: See ADN above
- ARINC 629: Commercial Aircraft (Boeing 777)
- ARINC 708: Weather Radar for Commercial Aircraft
- ARINC 717: Flight Data Recorder for Commercial Aircraft
- ARINC 825: CAN bus for commercial aircraft (for example Boeing 787 and Airbus A350)
- IEEE 1394b: Military Aircraft
- MIL-STD-1553: Military Aircraft
- MIL-STD-1760: Military Aircraft

- TTP – Time-Triggered Protocol: Boeing 787 Dreamliner, Airbus A380, Fly-By-Wire Actuation Platforms from Parker Aerospace

- TTEthernet – Time-Triggered Ethernet: NASA Orion Spacecraft

Guidance System

A guidance system is a virtual or physical device, or a group of devices implementing a guidance process used for controlling the movement of a ship, aircraft, missile, rocket, satellite, or any other moving object. Guidance is the process of calculating the changes in position, velocity, attitude, and/or rotation rates of a moving object required to follow a certain trajectory and/or attitude profile based on information about the object's state of motion.

A guidance system is usually part of a Guidance, navigation and control system, whereas navigation refers to the systems necessary to calculate the current position and orientation based on sensor data like those from compasses, GPS receivers, Loran-C, star trackers, inertial measurement units, altimeters, etc. The output of the navigation system, the navigation solution, is an input for the guidance system, among others like the environmental conditions (wind, water, temperature, etc.) and the vehicle's characteristics (i.e. mass, control system availability, control systems correlation to vector change, etc.). In general, the guidance system computes the instructions for the control system, which comprises the object's actuators (e.g., thrusters, reaction wheels, body flaps, etc.), which are capable to manipulate the flight path and orientation of the object without direct or continuous human control.

One of the earliest examples of a true guidance system is that used in the German V-1 during World War II. The navigation system consisted of a simple gyroscope, an airspeed sensor, and an altimeter. The guidance instructions were target altitude, target velocity, cruise time, engine cut off time.

A guidance system has three major sub-sections: Inputs, Processing, and Outputs. The input section includes sensors, course data, radio and satellite links, and other information sources. The processing section, composed of one or more CPUs, integrates this data and determines what actions, if any, are necessary to maintain or achieve a proper heading. This is then fed to the outputs which can directly affect the system's course. The outputs may control speed by interacting with devices such as turbines, and fuel pumps, or they may more directly alter course by actuating ailerons, rudders, or other devices.

History

Inertial guidance systems were originally developed for rockets. American rocket pioneer Robert Goddard experimented with rudimentary gyroscopic systems. Dr. Goddard's systems were of great interest to contemporary German pioneers including Wernher von Braun. The systems entered more widespread use with the advent of spacecraft, guided missiles, and commercial airliners.

US guidance history centers around 2 distinct communities. One driven out of Caltech and NASA Jet Propulsion Laboratory, the other from the German scientists that developed the early V2 rocket guidance and MIT. The GN&C system for V2 provided many innovations and was the most sophisticated military weapon in 1942 using self-contained closed loop guidance. Early V2s leveraged 2 gyroscopes and lateral accelerometer with a simple analog computer to adjust the azimuth for the rocket in flight. Analog computer signals were used to drive 4 external rudders on the tail fins for flight control. Von Braun engineered the surrender of 500 of his top rocket scientists, along with plans and test vehicles, to the Americans. They arrived in Fort Bliss, Texas in 1945 and were subsequently moved to Huntsville, Al in 1950 (aka Redstone arsenal). Von Braun's passion was interplanetary space flight. However his tremendous leadership skills and experience with the V-2 program made him invaluable to the US military. In 1955 the Redstone team was selected to put America's first satellite into orbit putting this group at the center of both military and commercial space.

The Jet Propulsion Laboratory traces its history from the 1930s, when Caltech professor Theodore von Karman conducted pioneering work in rocket propulsion. Funded by Army Ordnance in 1942, JPL's early efforts would eventually involve technologies beyond those of aerodynamics and propellant chemistry. The result of the Army Ordnance effort was JPL's answer to the German V-2 missile, named MGM-5 Corporal, first launched in May 1947. On December 3, 1958, two months after the National Aeronautics and Space Administration (NASA) was created by Congress, JPL was transferred from Army jurisdiction to that of this new civilian space agency. This shift was due to the creation of a military focused group derived from the German V2 team. Hence, beginning in 1958, NASA JPL and the Caltech crew became focused primarily on unmanned flight and shifted away from military applications with a few exceptions. The community surrounding JPL drove tremendous innovation in telecommunication, interplanetary exploration and earth monitoring (among other areas).

In the early 1950s, the US government wanted to insulate itself against over dependency on the Germany team for military applications. Among the areas that were domestically "developed" was missile guidance. In the early 1950s the MIT Instrumentation Laboratory (later to become the Charles Stark Draper Laboratory, Inc.) was chosen by the Air Force Western Development Division to provide a self-contained guidance system backup to Convair in San Diego for the new Atlas intercontinental ballistic missile. The technical monitor for the MIT task was a young engineer named Jim Fletcher who later served as the NASA Administrator. The Atlas guidance system was to be a combination of an on-board autonomous system, and a ground-based tracking and command system. This was the beginning of a philosophic controversy, which, in some areas, remains unresolved. The self-contained system finally prevailed in ballistic missile applications for obvious reasons. In space exploration, a mixture of the two remains.

In the summer of 1952, Dr. Richard Battin and Dr. J. Halcombe ("Hal") Laning Jr., researched computational based solutions to guidance as computing began to step out of the analog approach. As computers of that time were very slow (and missiles very fast) it was extremely important to develop programs that were very efficient. Dr. J. Halcombe Laning, with the help of Phil Hankins and Charlie Werner, initiated work on MAC, an algebraic programming language for the IBM 650, which was completed by early spring of 1958. MAC became the work-horse of the MIT lab. MAC is an extremely readable language having a three-line format, vector-matrix notations and mnemonic and indexed subscripts. Today's Space Shuttle (STS) language called HAL, (developed by Intermetrics, Inc.) is a direct offshoot of MAC. Since the principal architect of HAL was Jim Miller, who co-authored with Hal Laning a report on the MAC system, it is a reasonable speculation that the space shuttle language is named for Jim's old mentor, and not, as some have suggested, for the electronic superstar of the Arthur Clarke movie "2001-A Space Odyssey." (Richard Battin, AIAA 82-4075, April 1982)

Hal Laning and Richard Battin undertook the initial analytical work on the Atlas inertial guidance in 1954. Other key figures at Convair were Charlie Bossart, the Chief Engineer, and Walter Schweidetzky, head of the guidance group. Walter had worked with Wernher von Braun at Peenemuende during World War II.

The initial "Delta" guidance system assessed the difference in position from a reference trajectory. A velocity to be gained (VGO) calculation is made to correct the current trajectory with the objective of driving VGO to Zero. The mathematics of this approach were fundamentally valid, but dropped because of the challenges in accurate inertial navigation (e.g. IMU Accuracy) and analog computing power. The challenges faced by the "Delta" efforts were overcome by the "Q system" of guidance. The "Q" system's revolution was to bind the challenges of missile guidance (and associated equations of motion) in the matrix Q. The Q matrix represents the partial derivatives of the velocity with respect to the position vector. A key feature of this approach allowed for the components of the vector cross product (v, xdv,/dt) to be used as the basic autopilot rate signals-a technique that became known as "cross-product steering." The Q-system was presented at the first Technical Symposium on Ballistic Missiles held at the Ramo-Wooldridge Corporation in Los Angeles on June 21 and 22, 1956. The "Q System" was classified information through the 1960s. Derivations of this guidance are used for today's military missiles. The CSDL team remains a leader in the military guidance and is involved in projects for most divisions of the US military.

On August 10 of 1961 NASA Awarded MIT a contract for preliminary design study of a guidance and navigation system for Apollo program. Today's space shuttle guidance is named PEG4 (Powered Explicit Guidance). It takes into account both the Q system and the predictor-corrector attributes of the original "Delta" System (PEG

Guidance). Although many updates to the shuttles navigation system have taken place over the last 30 years (ex. GPS in the OI-22 build), the guidance core of today's Shuttle GN&C system has evolved little. Within a manned system, there is a human interface needed for the guidance system. As Astronauts are the customer for the system, many new teams are formed that touch GN&C as it is a primary interface to "fly" the vehicle. For the Apollo and STS (Shuttle system) CSDL "designed" the guidance, McDonnell Douglas wrote the requirements and IBM programmed the requirements.

Much system complexity within manned systems is driven by "redundancy management" and the support of multiple "abort" scenarios that provide for crew safety. Manned US Lunar and Interplanetary guidance systems leverage many of the same guidance innovations (described above) developed in the 1950s. So while the core mathematical construct of guidance has remained fairly constant, the facilities surrounding GN&C continue to evolve to support new vehicles, new missions and new hardware. The center of excellence for the manned guidance remains at MIT (CSDL) as well as the former McDonnell Douglas Space Systems (in Houston).

Guidance Systems

Guidance systems consist of 3 essential parts: navigation which tracks current location, guidance which leverages navigation data and target information to direct flight control "where to go", and control which accepts guidance commands to effect change in aerodynamic and/or engine controls.

Navigation is the art of determining where you are, a science that has seen tremendous focus in 1711 with the Longitude prize. Navigation aids either measure position from a *fixed* point of reference (ex. landmark, north star, LORAN Beacon), *relative* position to a target (ex. radar, infra-red, ...) or track *movement* from a known position/starting point (e.g. IMU). Today's complex systems use multiple approaches to determine current position. For example, today's most advanced navigation systems are embodied within the Anti-ballistic missile, the RIM-161 Standard Missile 3 leverages GPS, IMU and ground segment data in the boost phase and relative position data for intercept targeting. Complex systems typically have multiple redundancy to address drift, improve accuracy (ex. relative to a target) and address isolated system failure. Navigation systems therefore take multiple inputs from many different sensors, both internal to the system and/or external (ex. ground based update). Kalman filter provides the most common approach to combining navigation data (from multiple sensors) to resolve current position. Example navigation approaches:

- Celestial navigation is a position fixing technique that was devised to help sailors cross the featureless oceans without having to rely on dead reckoning to enable them to strike land. Celestial navigation uses angular measurements (sights) between the horizon and a common celestial object. The Sun is most often measured. Skilled navigators can use the Moon, planets or one of 57 navigational stars whose coordinates are tabulated in nautical almanacs. Historical tools include a sextant, watch and ephemeris data. Today's space shuttle, and most interplanetary spacecraft, use optical systems to calibrate inertial navigation systems: Crewman Optical Alignment Sight (COAS), Star Tracker.

- Inertial Measurement Units (IMUs) are the primary inertial system for maintaining current position (navigation) and orientation in missiles and aircraft. They are complex machines with one or more rotating Gyroscopes that can rotate freely in 3 degrees of motion within a complex gimbal system. IMUs are "spun up" and calibrated prior to launch. A minimum of 3 separate IMUs are in place within most complex systems. In addition to relative position, the IMUs contain accelerometers which can measure acceleration in all axes. The position data, combined with acceleration data provide the necessary inputs to "track" motion of a vehicle. IMUs have a tendency to "drift", due to friction and accuracy. Error correction to address this drift can be provided via ground link telemetry, GPS, radar, optical celestial navigation and other navigation aids. When targeting another (moving) vehicle, relative vectors become paramount. In this situation, navigation aids which provide updates of position *relative to the target* are more important. In addition to the current position, inertial navigation systems also typically estimate a predicted position for future computing cycles.

- Astro-inertial guidance is a sensor fusion/information fusion of the Inertial guidance and Celestial navigation.

- Long-range Navigation (LORAN) : This was the predecessor of GPS and was (and to an extent still is) used primarily in commercial sea transportation. The system works by triangulating the ship's position based on directional reference to known transmitters.

- Global Positioning System (GPS) : GPS was designed by the US military with the primary purpose of addressing "drift" within the inertial navigation of Submarine-launched ballistic missile(SLBMs) prior to launch. GPS transmits 2 signal types: military and a commercial. The accuracy of the military signal is classified but can be assumed to be well under 0.5 meters. GPS is a system of 24 satellites orbiting in unique planes 10.9-14.4 Nautical miles above the earth. The Satellites are in well defined orbits and transmit highly accurate time information which can be used to triangulate position.

- Radar/Infrared/Laser : This form of navigation provides information to guidance *relative to a known target*, it has both civilian (ex rendezvous) and military applications.

- o   active (employs own radar to illuminate the target),

- o   passive (detects target's radar emissions),

- o   semiactive radar homing,

- o   Infrared homing : This form of guidance is used exclusively for military munitions, specifically air-to-air and surface-to-air missiles. The missile's seeker head homes in on the infrared (heat) signature from the target's engines (hence the term "heat-seeking missile"),

- o   Ultraviolet homing, used in FIM-92 Stinger - more resistive to countermeasures, than IR homing system

- o   Laser guidance : A laser designator device calculates relative position to a highlighted target. Most are familiar with the military uses of the technology on Laser-guided bomb. The space shuttle crew leverages a hand held device to feed information into rendezvous planning. The primary limitation on this device is that it requires a line of sight between the target and the designator.

- o   Terrain contour matching (TERCOM). Uses a ground scanning radar to "match" topography against digital map data to fix current position. Used by cruise missiles such as the Tomahawk (missile).

Guidance is the "driver" of a vehicle. It takes input from the navigation system (where am I) and uses targeting information (where do I want to go) to send signals to the flight control system that will allow the vehicle to reach its destination (within the operating constraints of the vehicle). The "targets" for guidance systems are one or more state vectors (position and velocity) and can be inertial or relative. During powered flight, guidance is continually calculating steering directions for flight control. For example, the space shuttle targets an altitude, velocity vector, and gamma to drive main engine cut off. Similarly, an Intercontinental ballistic missile also targets a vector. The target vectors are developed to fulfill the mission and can be preplanned or dynamically created.

Control. Flight control is accomplished either aerodynamically or through powered controls such as engines. Guidance sends signals to flight control. A Digital Autopilot (DAP) is the interface between guidance and control. Guidance and the DAP are responsible for calculating the precise instruction for each flight control. The DAP provides feedback to guidance on the state of flight controls.

Electrically Powered Spacecraft Propulsion

An electrically powered spacecraft propulsion system uses electrical energy to change the velocity of a spacecraft. Most of these kinds of spacecraft propulsion systems work by electrically expelling propellant (reaction mass) at high speed, but electrodynamic tethers work by interacting with a planet's magnetic field.

6 kW Hall thruster in operation at the NASA Jet Propulsion Laboratory.

Electric thrusters typically use much less propellant than chemical rockets because they have a higher exhaust speed (operate at a higher specific impulse) than chemical rockets. Due to limited electric power the thrust is much weaker compared to chemical rockets, but electric propulsion can provide a small thrust for a long time. Electric propulsion can achieve high speeds over long periods and thus can work better than chemical rockets for some deep space missions.

Electric propulsion is now a mature and widely used technology on spacecraft. Russian satellites have used electric propulsion for decades. As of 2013, over 200 spacecraft operated throughout the solar system use electric propulsion for station keeping, orbit raising, or primary propulsion. In the future, the most advanced electric thrusters may be able to impart a Delta-v of 100 km/s, which is enough to take a spacecraft to the outer planets of the Solar System (with nuclear power), but is insufficient for interstellar travel. Also, an electro-rocket with an external power source (transmissible through laser on the solar panels) has a theoretical possibility for interstellar flight. However, electric propulsion is not a method suitable for launches from the Earth's surface, as the thrust for such systems is too weak.

History

The idea of electric propulsion for spacecraft dates back to 1911, introduced in a publication by Konstantin Tsiolkovsky. Earlier, Robert Goddard had noted such a possibility in his personal notebook.

The first in the world designed and tested electric propulsion was in 1929-1931 in Leningrad. Already in 1950 at the initiative of S.P. Korolev, I.V. Kurchatov and L.A. Artsimovich it adopted a program of research and development of various electrical rocket engines.

Electrically powered propulsion with a nuclear reactor was considered by Dr. Tony Martin for interstellar Project Daedalus in 1973, but the novel approach was rejected because of very low thrust, the heavy equipment needed to convert nuclear energy into electricity, and as a result a small acceleration, which would take a century to achieve the desired speed.

The demonstration of electric propulsion was an ion engine carried on board the SERT-1 (Space Electric Rocket Test) spacecraft, launched on 20 July 1964 and it operated for 31 minutes. A follow-up mission launched on 3 February 1970, SERT-2, carried two ion thrusters, one operated for more than five months and the other for almost three months.

By the early 2010s, many satellite manufacturers were offering electric propulsion options on their satellites—mostly for on-orbit attitude control—while some commercial communication satellite operators were beginning to use them for geosynchronous orbit insertion in place of traditional chemical rocket engines.

Types

Ion and Plasma Drives

This type of rocket-like reaction engine uses electric energy to obtain thrust from propellant carried with the vehicle. Unlike rocket engines, these kinds of engines do not necessarily have rocket nozzles, and thus many types are not considered true rockets.

Electric propulsion thrusters for spacecraft may be grouped in three families based on the type of force used to accelerate the ions of the plasma:

Electrostatic

If the acceleration is caused mainly by the Coulomb force (i.e. application of a static electric field in the direction of the acceleration) the device is considered electrostatic.

- Gridded ion thruster
    - NASA Solar Technology Application Readiness (NSTAR)
    - HiPEP
    - Radiofrequency ion thruster
- Hall effect thruster
    - SPT – Stationary Plasma Thruster
    - TAL – Thruster with Anode Layer
- Colloid ion thruster
- Field Emission Electric Propulsion

- Nano-particle field extraction thruster

- Contact ion thruster

- Plasma separator ion thruster

- Radioisotopic ion thruster

Electrothermal

The electrothermal category groups the devices where electromagnetic fields are used to generate a plasma to increase the temperature of the bulk propellant. The thermal energy imparted to the propellant gas is then converted into kinetic energy by a nozzle of either solid material or magnetic fields. Low molecular weight gases (e.g. hydrogen, helium, ammonia) are preferred propellants for this kind of system.

An electrothermal engine uses a nozzle to convert the heat of a gas into the linear motion of its molecules so it is a true rocket even though the energy producing the heat comes from an external source.

Performance of electrothermal systems in terms of specific impulse (Isp) is somewhat modest (500 to ~1000 seconds), but exceeds that of cold gas thrusters, monopropellant rockets, and even most bipropellant rockets. In the USSR, electrothermal engines were used since 1971; the Soviet "Meteor-3", "Meteor-Priroda", "Resurs-O" satellite series and the Russian "Elektro" satellite are equipped with them. Electrothermal systems by Aerojet (MR-510) are currently used on Lockheed Martin A2100 satellites using hydrazine as a propellant.

- Arcjet

- Microwave arcjet

- Resistojet

Electromagnetic

If ions are accelerated either by the Lorentz force or by the effect of an electromagnetic fields where the electric field is not in the direction of the acceleration, the device is considered electromagnetic.

- Electrodeless plasma thruster

- MPD thruster

- Pulsed inductive thruster

- Pulsed plasma thruster

- Helicon Double Layer Thruster

- Variable specific impulse magnetoplasma rocket (VASIMR)

Non-ion Drives

Photonic

Photonic drive does not expel matter for reaction thrust, only photons.

Electrodynamic Tether

Electrodynamic tethers are long conducting wires, such as one deployed from a tether satellite, which can operate on electromagnetic principles as generators, by converting their kinetic energy to electric energy, or as motors, converting electric energy to kinetic energy. Electric potential is generated across a conductive tether by its motion through the Earth's magnetic field. The choice of the metal conductor to be used in an electrodynamic tether is determined by a variety of factors. Primary factors usually include high electrical conductivity, and low density. Secondary factors, depending on the application, include cost, strength, and melting point.

Unconventional

The principle of action of these theoretical devices is not well explained by the currently-understood laws of physics.

- Quantum Vacuum Plasma Thruster

- EM Drive or Cannae Drive

Steady vs. Unsteady

Electric propulsion systems can also be characterized as either steady (continuous firing for a prescribed duration) or unsteady (pulsed firings accumulating to a desired impulse). However, these classifications are not unique to electric propulsion systems and can be applied to all types of propulsion engines.

Dynamic Properties

Electrically powered rocket engines provide lower thrust compared to chemical rockets by several orders of magnitude because of the limited electrical power possible to provide in a spacecraft. A chemical rocket imparts energy to the combustion products directly, whereas an electrical system requires several steps. However, the high velocity and lower reaction mass expended for the same thrust allows electric rockets to run for a long time. This differs from the typical chemical-powered spacecraft, where the engines run only in short intervals of time, while the spacecraft mostly follows an inertial trajectory. When near a planet, low-thrust propulsion may not offset the gravitational attraction of the planet. An electric rocket engine cannot provide enough thrust to lift the vehicle from a planet's surface, but a low thrust applied for a long interval can allow a spacecraft to maneuver near a planet.

Stabilizer (Aeronautics)

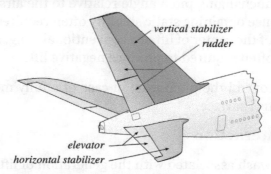

Vertical and horizontal stabilizer units on an Airbus A380 airliner

An aircraft stabilizer is an aerodynamic surface, typically including one or more movable control surfaces, that provides longitudinal (pitch) and/or directional (yaw) stability and control. A stabilizer can feature a fixed or adjustable structure on which any movable control surfaces are hinged, or it can itself be a fully movable surface such as a stabilator. Depending on the context, "stabilizer" may sometimes describe only the front part of the overall surface.

In the conventional aircraft configuration, separate vertical (fin) and horizontal (tailplane) stabilizers form an empennage positioned at the tail of the aircraft. Other arrangements of the empennage, such as the V-tail configuration, feature stabilizers which contribute to a combination of longitudinal and directional stabilization and control.

Longitudinal stability and control may be obtained with other wing configurations, including canard, tandem wing and tailless aircraft.

Some types of aircraft are stabilized with electronic flight control; in this case, fixed and movable surfaces located anywhere along the aircraft may serve as active motion dampers or stabilizers.

Horizontal Stabilizers

A longitudinal stabilizer is used to maintain the aircraft in longitudinal balance, or *trim*: it exerts a vertical force at a distance so that the summation of pitch moments about the center of gravity is zero. The vertical force exerted by the stabilizer to this effect varies with flight conditions, in particular according to the aircraft lift coefficient and wing flaps deflection which both affect the position of the center of lift, and with the position of the aircraft center of gravity (which changes with aircraft loading). Transonic flight makes special demands on horizontal stabilizers, since the crossing of the sound barrier is associated with a sudden move aft of the center of lift.

Another role of a longitudinal stabilizer is to provide longitudinal static stability. Stability can be defined only when the vehicle is in trim; it refers to the tendency of the aircraft to return to the trimmed condition if it is disturbed. This maintains a constant aircraft attitude, with unchanging pitch angle relative to the airstream, without active input from the pilot. Since obtaining static stability often requires that the aircraft center of gravity be ahead of the center of lift of a conventional wing, a stabilizer positioned aft of the wing is then often required to produce negative lift.

The elevator serve to control the pitch axis; in case of a fully movable tail, the entire assembly acts as a control surface.

Wing-stabilizer Interaction

The upwash and downwash associated with the generation of lift is the source of aerodynamic interaction between the wing and stabilizer, which translates into a change in the effective angle of attack for each surface. The influence of the wing on a tail is much more significant than the opposite effect and can be modeled using the Prandtl lifting-line theory; however, an accurate estimation of the interaction between multiple surfaces requires computer simulations or wind tunnel tests.

Horizontal Stabilizer Configurations

Conventional Tailplane

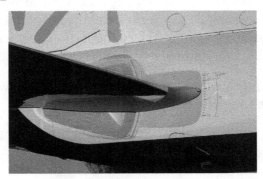

The adjustable horizontal stabilizer of an Embraer 170,
with markings showing nose-up and nose-down trim angles

In the conventional configuration the horizontal stabilizer is a small horizontal tail or tailplane located to the rear of the aircraft. This is the most common configuration.

On many aircraft, the tailplane assembly consists of a fixed surface fitted with a hinged aft elevator surface. Trim tabs may be used to relieve pilot input forces; conversely in some cases, such as small aircraft with all-moving stabilizers, anti-servo tabs are used to increase these forces.

Most airliners and transport aircraft feature a large, slow-moving *trimmable tail plane* which is combined with independently-moving elevators. The elevators are controlled by the pilot or autopilot and primarily serve to change the aircraft's attitude, while the whole assembly is used to trim (maintaining horizontal static equilibrium) and stabilize the aircraft in the pitch axis.

Many supersonic aircraft feature an all-moving tail assembly, also named stabilator, where the entire surface is adjustable.

Variants on the conventional configuration include the T-tail, Cruciform tail, Twin tail and Twin-boom mounted tail.

Three-surface Aircraft

The three-surface configuration of the Piaggio P-180 Avanti

Three-surface aircraft such as the Piaggio P.180 Avanti or the Scaled Composites Triumph and Catbird, the tailplane is a stabilizer as in conventional aircraft; the frontplane, called foreplane or canard, provides lift and serves as a balancing surface.

Some earlier three-surface aircraft, such as the Curtiss AEA June Bug or the Voisin 1907 biplane, were of conventional layout with an additional front pitch control surface which was called "elevator" or sometimes "stabilisateur". Lacking elevators, the tailplanes of these aircraft were not what is now called conventional stabilizers. For example, the Voisin was a tandem-lifting layout (main wing and rear wing) with a foreplane that was neither stabilizing nor mainly lifting; it was called an *"équilibreur"* ("balancer"), and used as a pitch control and trim surface.

Canard Aircraft

In the canard configuration, a small wing, or *foreplane*, is located in front of the main wing. Some authors call it a stabilizer or give to the foreplane alone a stabilizing role, although as far as pitch stability is concerned, a foreplane is generally described as a destabilizing surface, the main wing providing the stabilizing moment in pitch.

The canard configuration of the Beechcraft Starship

In naturally unstable aircraft, the canard surfaces may be used as an active part of the artificial stability system, and are sometimes named horizontal stabilizers.

Tailless Aircraft

The tailless configuration of Concorde

Tailless aircraft lack a separate horizontal stabilizer. In a tailless aircraft, the horizontal stabilizing surface is part of the main wing. Longitudinal stability in tailless aircraft is achieved by designing the aircraft so that its aerodynamic center is behind the center of gravity. This is generally done by modifying the wing design, for example by varying the angle of incidence in the span-wise direction (wing washout or twist), or by using reflexed camber airfoils.

Vertical Stabilizers

A vertical stabilizer provides directional (or yaw) stability and usually comprises a fixed *fin* and movable control *rudder* hinged to its rear edge. Less commonly, there is no hinge and the whole fin surface is pivoted for both stability and control.

When an aircraft encounters a horizontal gust of wind, yaw stability causes the aircraft to turn into the wind, rather than turn in the same direction.

Fuselage geometry, engine nacelles and rotating propellers all influence lateral static stability and affect the required size of the stabilizer.

Tailless Directional Stabilization

Although the use of a vertical stabilizer is most common, it is possible to obtain directional stability with no discrete vertical stabilizer.

One approach is to use differential drag. By increasing drag on the outer wing and reducing drag on the inner wing, a corrective moment can be applied to restore the aircraft attitude and correct the yaw.

This occurs when the wing is swept back and in some cases, as for example on the Rogallo wing often used for hang gliders, means that no fin is needed. When the aircraft is rotated in yaw the outer wing sweep is reduced, so increasing drag, while the inner wing sweep increases, reducing drag. This change in the drag distribution creates a restoring moment.

Another approach is to use differential air braking to affect the drag directly. This technique is suited to Electronic flight controls, as on the Northrop Grumman B-2 flying wing.

Combined Longitudinal - Directional Stabilizers

The Beechcraft Bonanza, the most common example of V-tail empennage configuration

On some aircraft, horizontal and vertical stabilizers are combined in a pair of surfaces named V-tail. In this arrangement, two stabilizers (fins and rudders) are mounted at 90 - 120° to each other, giving a larger horizontal projected area than vertical one as in the majority of conventional tails. The moving control surfaces are then named *rudder-vators*. The V-tail thus acts both as a yaw and pitch stabilizer.

Although it may seem that the V-tail configuration can result in a significant reduction of the tail wetted area, it suffers from an increase in control-actuation complexity, as well as complex and detrimental aerodynamic interaction between the two surfaces.

This often results in an upsizing in the total area that reduces or negates the original benefit. The Beechcraft Bonanza light aircraft was originally designed with a V-tail.

Others combined layouts exist. The General Atomics MQ-1 Predator unmanned aircraft has an *inverted V-tail*. The tail surfaces of the Lockheed XFV could be described as a V-tail with surfaces that extended through the fuselage to the opposite side. The LearAvia Lear Fan had a *Y-tail*. All twin tail arrangements with a tail dihedral angle will provide a combination of longitudinal and directional stabilization.

Vertical Stabilizer

The vertical stabilizers, vertical stabilisers, or fins, of aircraft, missiles or bombs are typically found on the aft end of the fuselage or body, and are intended to reduce aerodynamic side slip and provide direction stability. It is analogous to a skeg on boats and ships.

Boeing B-29 Superfortress showing conventional single vertical stabilizer

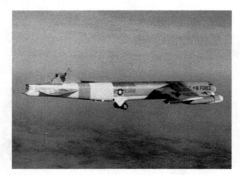

A Boeing B-52 with its vertical stabilizer ripped off. Despite the catastrophic failure of the stabilizer, the plane managed to land safely.

On aircraft, vertical stabilizers generally point upwards. These are also known as the vertical tail, and are part of an aircraft's empennage. This upright mounting position has two major benefits: The drag of the stabilizer increases at speed which creates a nose-up moment that help to slow down the aircraft that prevent dangerous overspeed, and when the aircraft banks the stabilizer produce lift which counter the banking moment and keep the aircraft upright at the absence of control input. If the vertical stabilizer was mounted on the underside, it would produce a positive feedback whenever the aircraft dive or bank, which is inherently unstable. The trailing end of the stabilizer is typically movable, and called the rudder; this allows the aircraft pilot to control yaw.

Often navigational radio or airband transceiver antennas are placed on or inside the vertical tail. In all known trijets (aircraft with 3 engines), the vertical stabilizer houses the central engine or engine inlet duct.

Vertical stabilizers, or fins, have also been used in automobiles, specifically in top level motor sports, with the concept making a resurgence in both Formula 1 and Le Mans Prototype racing.

A few aircraft models have a ventral fin under the rear end. Normally this is small, or can fold sideways, to allow landing. Both the North American X-15 supersonic/hypersonic experimental aircraft, and the late World War II German twin-engined Dornier Do 335 heavy fighter used differing forms of the cruciform tail stabilizing surface format.

Types

Single

Conventional Tail

The vertical stabilizer is mounted exactly vertically, and the horizontal stabilizer is directly mounted to the empennage (the rear fuselage). This is the most common vertical stabilizer configuration.

The conventional tail of an Airbus A380, with the vertical stabilizer exactly vertical

Tails of Iberia aircraft at Madrid Barajas Airport.

T-tail

A T-tail has the horizontal stabilizer mounted at the top of the vertical stabilizer. It is commonly seen on rear-engine aircraft, such as the Bombardier CRJ200, the Fokker 70, the Boeing 727, the Vickers VC10 and Douglas DC-9, and most high performance gliders.

T-tails are often incorporated on configurations with fuselage mounted engines to keep the horizontal stabilizer away from the engine exhaust plume.

T-tail aircraft are more susceptible to pitch-up at high angles of attack. This pitch-up results from a reduction in the horizontal stabilizer's lifting capability as it passes through the wake of the wing at moderate angles of attack. This can also result in a deep stall condition.

T-tails present structural challenges since loads on the horizontal stabilizer must be transmitted through the vertical tail.

Cruciform Tail

The cruciform tail is arranged like a cross, the most common configuration having the horizontal stabilizer intersecting the vertical tail somewhere near the middle. The PBY Catalina uses this configuration. The "push-pull" twin engined Dornier Do 335 World War II German fighter used a cruciform tail consisting of four separate surfaces, arranged in dorsal, ventral, and both horizontal locations, to form its cruciform tail, just forward of the rear propeller.

Falconjets from Dassault always have cruciform tail.

Multiple Stabilizers

Twin Tail

The twin tail of a Chrislea Super Ace, built in 1948

Rather than a single vertical stabilizer, a twin tail has two. These are vertically arranged, and intersect or are mounted to the ends of the horizontal stabilizer. The

Beechcraft Model 18 and many modern military aircraft such as the American F-14, F-15, and F/A-18 use this configuration. The F/A-18, F-22 Raptor, and F-35 Lightning II have tailfins that are canted outward, to the point that they have some authority as horizontal control surfaces; both aircraft are designed to deflect their rudders inward during takeoff to increase pitching moment. A twin tail may be either H-tail, twin fin/rudder construction attached to a single fuselage such as North American B-25 Mitchell or Avro Lancaster, or twin boom tail, the rear airframe consisting of two separate fuselages each sporting one single fin/rudder, such as Lockheed P-38 Lightning or C-119 Boxcar.

Triple Tail

A Lockheed Constellation with a triple tail

A variation on the twin tail, it has three vertical stabilizers. An example of this configuration is the Lockheed Constellation. On the Constellation it was done to give the airplane maximum vertical stabilizer area while keeping the overall height low enough so that it could fit into maintenance hangars.

V-tail

A V-tail has no distinct vertical or horizontal stabilizers. Rather, they are merged into control surfaces known as ruddervators which control both pitch and yaw. The arrangement looks like the letter V, and is also known as a *butterfly tail*. The Beechcraft Bonanza Model 35 uses this configuration, as does the F-117 Nighthawk, and many of Richard Schreder's HP series of homebuilt gliders.

Winglet

Winglets served double duty on Burt Rutan's canard pusher configuration VariEze and Long-EZ, acting as both a wingtip device and a vertical stabilizer. Several other derivatives of these and other similar aircraft use this design element.

Fins

The vertical stabilizer often employs a small fillet or "dorsal fin" at its forward base which helps to increase the stall angle of the vertical surface (thanks to vortex lift) and to prevent a phenomenon called rudder lock or rudder reversal. Rudder lock occurs when the force on a deflected rudder (in a steady sideslip) suddenly reverses as the vertical stabilizer stalls. This may leave the rudder stuck at full deflection with the pilot unable to recenter it. The fillet is sometimes called a dorsal fin.

Automotive/Motorsports Use

While vertical stabilizers have also been used in some race cars, such as the 1955 Jaguar D-type, the concept has seen sparing use until recently when the concept has seen a resurgence in Formula 1 and Le Mans endurance racing. The ostensible purpose of this is primarily to reduce sudden high speed yaw induced blow overs that would cause the cars to flip due to aerodynamic lift when subject to extreme yaw angles during cornering or in a spin. In addition to this, some Formula 1 teams utilized the wing as a way to disrupt the airflow to the rear wing reducing drag, the most radical system being the "F-duct" found in the MP4-25 (and later copied by Ferrari in the Ferrari F10), which could divert air from a duct in the front of the car, on demand by the driver, through a tunnel in the vertical fin onto the rear wing to stall it and reduce drag on the straights on which downforce wasn't needed. The system has since been banned for the 2011 Formula 1 season. For Le Mans Prototypes, the vertical stabilizer, dubbed the "Big Honking Fin" by some fans has become mandatory for all newly homologated sports prototypes.

Ferrari F10 with large rear vertical fin sprouting out of the airbox and leading into the rear wing.

Stealth Technology

Stealth technology also termed LO technology (low observable technology) is a sub-discipline of military tactics and passive electronic countermeasures, which cover a range of techniques used with personnel, aircraft, ships, submarines, missiles and satellites to make them less visible (ideally invisible) to radar, infrared, sonar and other detection methods. It corresponds to military camouflage for these parts of the electromagnetic spectrum (Multi-spectral camouflage).

F-117 stealth attack plane

Surcouf French stealth frigate

Development of modern stealth technologies in the United States began in 1958, where earlier attempts in preventing radar tracking of its U-2 spy planes during the Cold War by the Soviet Union had been unsuccessful. Designers turned to develop a particular shape for planes that tended to reduce detection, by redirecting electromagnetic waves from radars. Radar-absorbent material was also tested and made to reduce or block radar signals that reflect off from the surface of planes. Such changes to shape and surface composition form stealth technology as currently used on the Northrop Grumman B-2 Spirit "Stealth Bomber".

The concept of stealth is to operate or hide without giving enemy forces any indications as to the presence of friendly forces. This concept was first explored through camouflage by blending into the background visual clutter. As the potency of detection and interception technologies (radar, Infra-red search and track, surface-to-air missiles, etc.) have increased over time, so too has the extent to which the design and operation of military personnel and vehicles have been affected in response. Some military uniforms are treated with chemicals to reduce their infrared signature. A modern "stealth" vehicle is designed from the outset to have a chosen spectral signature. The degree of stealth embodied in a particular design is chosen according to the predicted capabilities of projected threats.

History

The concept of camouflage is known to predate warfare itself. Hunters have been using vegetation to conceal themselves perhaps as long as people have been hunting. In England, irregular units of gamekeepers in the 17th century were the first to adopt drab colours (common in 16th century Irish units) as a form of camouflage, following examples from the continent.

During World War I, the Germans experimented with the use of *Cellon* (Cellulose acetate), a transparent covering material, in an attempt to reduce the visibility of military aircraft. Single examples of the Fokker E.III *Eindecker* fighter monoplane, the Albatros C.I two-seat observation biplane, and the Linke-Hofmann R.I prototype heavy bomber were covered with *Cellon*. In fact, sunlight glinting from the material made the aircraft even more visible. *Celon* was also found to be quickly degraded both by sunlight and in-flight temperature changes so the attempt to make transparent aircraft was not proceeded with.

In 1916, the British modified a small SS class airship for the purpose of night-time reconnaissance over German lines on the Western Front. Fitted with a silenced engine and a black gas bag, the craft was both invisible and inaudible from the ground but several night-time flights over German-held territory produced little useful intelligence and the idea was dropped.

Diffused lighting camouflage, a shipborne form of counter-illumination camouflage, was trialled by the Royal Canadian Navy from 1941 to 1943. The concept was followed up, but for aircraft, by the Americans and the British: in 1945 a Grumman Avenger with Yehudi lights, reached 3,000 yards (2,700 m) from a ship before being sighted. This ability was rendered obsolete by radar.

The U-boat *U-480* may have been the first stealth submarine. It featured an anechoic tile rubber coating, one layer of which contained circular air pockets to defeat ASDIC sonar. Radar absorbent rubber/semiconductor composite paints and materials (codenames: "Sumpf", "Schornsteinfeger") were used by the Kriegsmarine on submarines in World War II. Tests showed they were effective in reducing radar signatures at both short (centimetres) and long (1.5 metre) wavelengths.

In 1960, the first stealth technology development program was initiated by USAF, by reducing the radar-cross-section of a Ryan Q-2C Firebee drone. This was achieved through specially designed screens over the air intake, radar-absorbent material on the fuselage and a special radar-absorbing paint.

In 1958, the U.S. Central Intelligence Agency requested funding for a reconnaissance aircraft to replace the existing U-2 spy planes, and Lockheed secured contractual rights to produce it. "Kelly" Johnson and his team at Lockheed's Skunk Works were assigned to produce the A-12 (or OXCART), the first of the previously top secret Blackbird series, which operated at high altitude of 70,000 to 80,000 ft and speed of Mach 3.2 to avoid radar detection. Radar absorbent material was used on U-2 spy planes, and various plane shapes designed to reduce radar detection were developed in earlier prototypes, named A1 to A11. In 1964, an optimal plane shape taking into account compactness was developed for another "Blackbird", the Lockheed SR-71. This aircraft surpassed prior models in both altitude (90,000 ft) and speed (Mach 3.3). The SR-71 included a number of stealthy features, notably its canted vertical stabilizers, the use of composite materials in key locations, and the overall finish in radar absorbing paint.

During the 1970s the U.S. Department of Defense launched project Lockheed Have Blue, with the aim of developing a stealth fighter. There was fierce bidding between Lockheed and Northrop to secure the multibillion-dollar contract. Lockheed incorporated into its bid a text written by the Soviet/Russian physicist Pyotr Ufimtsev from 1962, titled *Method of Edge Waves in the Physical Theory of Diffraction*, Soviet Radio, Moscow, 1962. In 1971 this book was translated into English with the same title by U.S. Air Force, Foreign Technology Division. The theory played a critical role in the design of American stealth-aircraft F-117 and B-2. Equations outlined in the paper quantified how a plane's shape would affect its detectability by radar, its radar cross-section (RCS). This was applied by Lockheed in computer simulation to design a novel shape they called the "Hopeless Diamond", a wordplay on the Hope Diamond, securing contractual rights to produce the F-117 Nighthawk starting in 1975. In 1977 Lockheed produced two 60% scale models under the Have Blue contract. The Have Blue program was a stealth technology demonstrator that lasted from 1976 to 1979. Also the Northrop Grumman Tacit Blue played a part in the development of composite material and curvilinear surfaces, as well as Low Observables, fly-by-wire, and other stealth technology innovations. The success of Have Blue led the Air Force to create the *Senior Trend* program which developed the F-117.

Principles

Stealth technology (or LO for "low observability") is not a single technology. It is a combination of technologies that attempt to greatly reduce the distances at which a person or vehicle can be detected; in particular radar cross section reductions, but also acoustic, thermal, and other aspects:

Radar Cross-section (RCS) Reductions

Almost since the invention of radar, various methods have been tried to minimize detection. Rapid development of radar during World War II led to equally rapid development of numerous counter radar measures during the period; a notable example of this was the use of chaff. Modern methods include Radar jamming and deception.

The term "stealth" in reference to reduced radar signature aircraft became popular during the late eighties when the Lockheed Martin F-117 stealth fighter became widely known. The first large scale (and public) use of the F-117 was during the Gulf War in 1991. However, F-117A stealth fighters were used for the first time in combat during Operation Just Cause, the United States invasion of Panama in 1989. Increased awareness of stealth vehicles and the technologies behind them is prompting the development of means to detect stealth vehicles, such as passive radar arrays and low-frequency radars. Many countries nevertheless continue to develop low-RCS vehicles because they offer advantages in detection range reduction and amplify the effectiveness of on-board systems against active radar homing threats.

Vehicle Shape

Aircraft

The F-35 Lightning II offers better stealthy features (such as this landing gear door) than prior American multi-role fighters, such as the F-16 Fighting Falcon

The possibility of designing aircraft in such a manner as to reduce their radar cross-section was recognized in the late 1930s, when the first radar tracking systems were employed, and it has been known since at least the 1960s that aircraft shape makes a significant difference in detectability. The Avro Vulcan, a British bomber of the 1960s, had a remarkably small appearance on radar despite its large size, and occasionally disappeared from radar screens entirely. It is now known that it had a fortuitously stealthy shape apart from the vertical element of the tail. Despite being designed before a low radar cross-section (RCS) and other stealth factors were ever a consideration, a Royal Aircraft Establishment technical note of 1957 stated that of all the aircraft so far studied, the Vulcan appeared by far the simplest radar echoing object, due to its shape: only one or two components contributing significantly to the echo at any aspect, compared with three or more on most other types. While writing about radar systems, authors Simon Kingsley and Shaun Quegan singled out the Vulcan's shape as acting to reduce the RCS. In contrast, the Tupolev 95 Russian long-range bomber (NATO reporting name 'Bear') was conspicuous on radar. It is now known that propellers and jet turbine blades produce a bright radar image; the Bear has four pairs of large (5.6 meter diameter) contra-rotating propellers.

Another important factor is internal construction. Some stealth aircraft have skin that is radar transparent or absorbing, behind which are structures termed re-entrant triangles. Radar waves penetrating the skin get trapped in these structures, reflecting off the internal faces and losing energy. This method was first used on the Blackbird series (A-12/YF-12A/Lockheed SR-71 Blackbird).

The most efficient way to reflect radar waves back to the emitting radar is with orthogonal metal plates, forming a corner reflector consisting of either a dihedral (two plates) or a trihedral (three orthogonal plates). This configuration occurs in the tail of a conventional aircraft, where the vertical and horizontal components of the tail are set at right angles. Stealth aircraft such as the F-117 use a different arrangement, tilting the tail surfaces to reduce corner reflections formed between them. A more radical method is to eliminate the tail completely, as in the B-2 Spirit. The B-2's clean, low-drag flying wing configuration not only gives it exceptional range but also reduces its radar profile. The flying wing design most closely resembles a so-called infinite flat plate (as vertical control surfaces dramatically increase RCS), the perfect stealth shape, as it would have no angles to reflect back radar waves.

In addition to altering the tail, stealth design must bury the engines within the wing or fuselage, or in some cases where stealth is applied to an extant aircraft, install baffles in the air intakes, so that the compressor blades are not visible to radar. A stealthy shape must be devoid of complex bumps or protrusions of any kind, meaning that weapons, fuel tanks, and other stores must not be carried externally. Any stealthy vehicle becomes un-stealthy when a door or hatch opens.

Parallel alignment of edges or even surfaces is also often used in stealth designs. The technique involves using a small number of edge orientations in the shape of the structure. For example, on the F-22A Raptor, the leading edges of the wing and the tail planes are set at the same angle. Other smaller structures, such as the air intake bypass doors and the air refueling aperture, also use the same angles. The effect of this is to return a narrow radar signal in a very specific direction away from the radar emitter rather than returning a diffuse signal detectable at many angles. The effect is sometimes called "glitter" after the very brief signal seen when the reflected beam passes across a detector. It can be difficult for the radar operator to distinguish between a glitter event and a digital glitch in the processing system.

Stealth airframes sometimes display distinctive serrations on some exposed edges, such as the engine ports. The YF-23 has such serrations on the exhaust ports. This is another example in the parallel alignment of features, this time on the external airframe.

Shaping requirements detract greatly from an aircraft's aerodynamic properties. The F-117 has poor aerodynamics, is inherently unstable, and cannot be flown without a fly-by-wire control system.

Similarly, coating the cockpit canopy with a thin film transparent conductor (vapor-deposited gold or indium tin oxide) helps to reduce the aircraft's radar profile, because radar waves would normally enter the cockpit, reflect off objects (the inside of a cockpit has a complex shape, with a pilot helmet alone forming a sizeable return), and possibly return to the radar, but the conductive coating creates a controlled shape that deflects the incoming radar waves away from the radar. The coating is thin enough that it has no adverse effect on pilot vision.

K32 HMS *Helsingborg*, a stealth ship

Ships

Ships have also adopted similar methods. Though the earlier Arleigh Burke-class destroyer incorporated some signature-reduction features., the Skjold-class corvette was the first coastal defence and the French La Fayette-class frigate the first ocean-going stealth ship to enter service. Other examples are the German Sachsen-class frigates, the Swedish Visby-class corvette, the USS *San Antonio* amphibious transport dock, and most modern warship designs.

Materials

Non-metallic Airframe

Dielectric composites are more transparent to radar, whereas electrically conductive materials such as metals and carbon fibers reflect electromagnetic energy incident on the material's surface. Composites may also contain ferrites to optimize the dielectric and magnetic properties of a material for its application.

Radar-absorbing Material

Radar-absorbent material (RAM), often as paints, are used especially on the edges of metal surfaces. While the material and thickness of RAM coatings can vary, the way they work is the same: absorb radiated energy from a ground or air based radar station into the coating and convert it to heat rather than reflect it back. Current technologies include dielectric composites and metal fibers containing ferrite isotopes. Paint comprises depositing pyramid like colonies on the reflecting superficies with the gaps filled with ferrite-based RAM. The pyramidal structure deflects the incident radar energy in the maze of RAM. A commonly used material is known as "Iron Ball Paint". Iron ball paint contains microscopic iron spheres that resonate in tune with incoming radio waves and dissipate the majority of their energy as heat, leaving little to bounce back to detectors. FSS are planar periodic structures that behave like filters to electromagnetic energy. The considered frequency selective surfaces are composed of conducting patch elements pasted on the ferrite layer. FSS are used for filtration and microwave absorption.

Radar Stealth Countermeasures and Limits

Low-frequency Radar

Shaping offers far fewer stealth advantages against low-frequency radar. If the radar wavelength is roughly twice the size of the target, a half-wave resonance effect can still generate a significant return. However, low-frequency radar is limited by lack of available frequencies (many are heavily used by other systems), by lack of accuracy of the diffraction-limited systems given their long wavelengths, and by the radar's size, making it difficult to transport. A long-wave radar may detect a target and roughly locate it, but not provide enough information to identify it, target it with weapons, or even to guide a fighter to it. Noise poses another problem, but that can be efficiently addressed using modern computer technology; Chinese "Nantsin" radar and many older Soviet-made long-range radars have been modified by supporting them with modern computers.

Multiple Emitters

Much of the stealth comes in in directions different than a direct return. Thus, detection can be better achieved if emitters are separate from receivers. One emitter separate from one receiver is termed bistatic radar; one or more emitters separate from more than one receiver is termed multistatic radar. Proposals exist to use reflections from emitters such as civilian radio transmitters, including cellular telephone radio towers.

Moore's Law

By Moore's law the processing power behind radar systems is rising over time. This will erode the ability of physical stealth to hide vehicles.

Ship's Wakes and Spray

Synthetic Aperture sidescan radars can be used to detect the location and heading of ships from their wake patterns. These may be detectable from orbit. When a ship moves through a seaway it throws up a cloud of spray which can be detected by radar.

Schlieren Signature

Anything that disturbs the atmosphere may be detected (Schlieren photography) because of the Schlieren effect caused by that atmospheric disturbance. This type of Measurement and signature intelligence detection falls under the category of Electro-optical MASINT.

Acoustics

Acoustic stealth plays a primary role in submarine stealth as well as for ground vehicles. Submarines use extensive rubber mountings to isolate and avoid mechanical noises that could reveal locations to underwater passive sonar arrays.

Early stealth observation aircraft used slow-turning propellers to avoid being heard by enemy troops below. Stealth aircraft that stay subsonic can avoid being tracked by sonic boom. The presence of supersonic and jet-powered stealth aircraft such as the SR-71 Blackbird indicates that acoustic signature is not always a major driver in aircraft design, as the Blackbird relied more on its extremely high speed and altitude.

One possible technique for reducing helicopter rotor noise is 'modulated blade spacing'. Standard rotor blades are evenly spaced, and produce greater noise at a particular frequency and its harmonics. Using varying degrees of spacing between the blades spreads the noise or acoustic signature of the rotor over a greater range of frequencies.

Visibility

The simplest technology is visual camouflage; the use of paint or other materials to color and break up the lines of the vehicle or person.

Most stealth aircraft use matte paint and dark colors, and operate only at night. Lately, interest in daylight Stealth (especially by the USAF) has emphasized the use of gray paint in disruptive schemes, and it is assumed that Yehudi lights could be used in the future to mask shadows in the airframe (in daylight, against the clear background of the sky, dark tones are easier to detect than light ones) or as a sort of active camouflage. The original B-2 design had wing tanks for a contrail-inhibiting chemical, alleged by some to be chlorofluorosulfonic acid, but this was replaced in the final design with a contrail sensor that alerts the pilot when he should change altitude and mission planning also considers altitudes where the probability of their formation is minimized.

In space, mirrored surfaces can be employed to reflect views of empty space toward known or suspected observers; this approach is compatible with several radar stealth schemes. Careful control of the orientation of the satellite relative to the observers is essential, and mistakes can lead to detectability enhancement rather than the desired reduction.

Infrared

An exhaust plume contributes a significant infrared signature. One means to reduce IR signature is to have a non-circular tail pipe (a slit shape) to minimize the exhaust cross sectional area and maximize the mixing of hot exhaust with cool ambient air. Often, cool air is deliberately injected into the exhaust flow to boost this process. According to the Stefan–Boltzmann law, this results in less energy (Thermal radiation in infrared spectrum) being released and thus reduces the heat signature. Sometimes, the jet exhaust is vented above the wing surface to shield it from observers below, as in the Lockheed F-117 Nighthawk, and the unstealthy Fairchild Republic A-10 Thunderbolt II. To achieve infrared stealth, the exhaust gas is cooled to the temperatures where the brightest wavelengths it radiates are absorbed by atmospheric carbon dioxide and water vapor, dramatically reducing the infrared visibility of the exhaust plume. Another way to reduce the exhaust temperature is to circulate coolant fluids such as fuel inside the exhaust pipe, where the fuel tanks serve as heat sinks cooled by the flow of air along the wings.

Ground combat includes the use of both active and passive infrared sensors and so the USMC ground combat uniform requirements document specifies infrared reflective quality standards.

Reducing Radio Frequency (RF) emissions

In addition to reducing infrared and acoustic emissions, a stealth vehicle must avoid radiating any other detectable energy, such as from onboard radars, communications systems, or RF leakage from electronics enclosures. The F-117 uses passive infrared and low light level television sensor systems to aim its weapons and the F-22 Raptor has an advanced LPI radar which can illuminate enemy aircraft without triggering a radar warning receiver response.

Measuring

The size of a target's image on radar is measured by the radar cross section or RCS, often represented by the symbol σ and expressed in square meters. This does not equal geometric area. A perfectly conducting sphere of projected cross sectional area 1 m^2 (i.e. a diameter of 1.13 m) will have an RCS of 1 m^2. Note that for radar wavelengths much less than the diameter of the sphere, RCS is independent of frequency. Conversely, a square flat plate of area 1 m^2 will have an RCS of $\sigma = 4\pi A^2 / \lambda^2$ (where A=area, λ=wavelength), or 13,982 m^2 at 10 GHz if the radar is perpendicular to the flat surface. At off-normal incident angles, energy is reflected away from the receiver, reducing the RCS. Modern stealth aircraft are said to have an RCS comparable with small birds or large insects, though this varies widely depending on aircraft and radar.

If the RCS was directly related to the target's cross-sectional area, the only way to reduce it would be to make the physical profile smaller. Rather, by reflecting much of the radiation away or by absorbing it, the target achieves a smaller radar cross section.

Tactics

Stealthy strike aircraft such as the Lockheed F-117 Nighthawk, designed by the famous Skunk Works, are usually used against heavily defended enemy sites such as Command and control centers or surface-to-air missile (SAM) batteries. Enemy radar will cover the airspace around these sites with overlapping coverage, making undetected entry by conventional aircraft nearly impossible. Stealthy aircraft can also be detected, but only at short ranges around the radars; for a stealthy aircraft there are substantial gaps in the radar coverage. Thus a stealthy aircraft flying an appropriate route can remain undetected by radar. Many ground-based radars exploit Doppler filter to improve sensitivity to objects having a radial velocity component with respect to the radar. Mission planners use their knowledge of enemy radar locations and the RCS pattern of the aircraft to design a flight path that minimizes radial speed while presenting the lowest-RCS aspects of the aircraft to the threat radar. To be able to fly these "safe" routes, it is necessary to understand an enemy's radar coverage. Airborne or mobile radar systems such as AWACS can complicate tactical strategy for stealth operation.

Research

Negative index metamaterials are artificial structures for which refractive index has a negative value for some frequency range, such as in microwave, infrared, or possibly optical. These offer another way to reduce detectability, and may provide electromagnetic near-invisibility in designed wavelengths.

Plasma stealth is a phenomenon proposed to use ionized gas (plasma) to reduce RCS of vehicles. Interactions between electromagnetic radiation and ionized gas have been studied extensively for many purposes, including concealing vehicles from radar. Various methods might form a layer or cloud of plasma around a vehicle to deflect or absorb radar, from simpler electrostatic to RF more complex laser discharges, but these may be difficult in practice.

Several technology research and development efforts exist to integrate the functions of aircraft flight control systems such as ailerons, elevators, elevons, flaps, and flaperons into wings to perform the aerodynamic purpose with the advantages of lower RCS for stealth via simpler geometries and lower complexity (mechanically simpler, fewer or no moving parts or surfaces, less maintenance), and lower mass, cost (up to 50% less), drag (up to 15% less during use) and, inertia (for faster, stronger control response to change vehicle orientation to reduce detection). Two promising approaches are flexible wings, and fluidics.

In flexible wings, much or all of a wing surface can change shape in flight to deflect air flow. Adaptive compliant wings are a military and commercial effort. The X-53 Active Aeroelastic Wing was a US Air Force, Boeing, and NASA effort.

In fluidics, fluid injection is being researched for use in aircraft to control direction, in two ways: circulation control and thrust vectoring. In both, larger more complex mechanical parts are replaced by smaller, simpler fluidic systems, in which larger forces in fluids are diverted by smaller jets or flows of fluid intermittently, to change the direction of vehicles.

In circulation control, near the trailing edges of wings, aircraft flight control systems are replaced by slots which emit fluid flows.

In thrust vectoring, in jet engine nozzles, swiveling parts are replaced by slots which inject fluid flows into jets to divert thrust. Tests show that air forced into a jet engine exhaust stream can deflect thrust up to 15 degrees. The U.S. FAA has conducted a study about civilizing 3D military thrust vectoring to help jetliners avoid crashes. According to this study, 65% of all air crashes can be prevented by deploying thrust vectoring means.

List of Stealth Aircraft

- F-22 Raptor
- SR-71 Blackbird
- F-117 Nighthawk
- F-35 Lightning II
- B-2 Spirit
- T-50 PAK FA

References

- Haddow, G.W.; Peter M. Grosz (1988). The German Giants - The German R-Planes 1914-1918 (3rd ed.). London: Putnam. ISBN 0-85177-812-7.

- Knott, E.F; Shaeffer, J.F.; Tuley, M.T. (2004). Radar cross section - Second Edition. Raleigh, North Carolina: SciTech Publishing. pp. 209–214. ISBN 1-891121-25-1. Retrieved 7 October 2009.

- Knott, Eugene; Shaeffer, John; Tuley, Michael (1993). Radar Cross Section, 2nd ed. Artech House, Inc. p. 231. ISBN 0-89006-618-3.

- "Vectored Propulsion, Supermanoeuvreability, and Robot Aircraft", by Benjamin Gal-Or, Springer Verlag, 1990, ISBN 0-387-97161-0, 3-540-97161-0.

- "DDG-51 Arleigh Burke-class". FAS website. Federation of American Scientists. Archived from the original on 24 December 2013. Retrieved 2 February 2011.

- "Stealth Helicopter: MH-X Advanced Special Operations Helicopter". GlobalSecurity.org. Retrieved 28 April 2012.

- "Naval Museum of Quebec". Diffused Lighting and its use in the Chaleur Bay. Royal Canadian Navy. Retrieved 18 September 2012.

- Hepcke, Gerhard (2007). "The Radar War, 1930-1945" (PDF). English translation by Hannah Liebmann. Radar World: 45. Retrieved 19 September 2012.

- Benson, Robert (November 1998). "The Arleigh Burke: Linchpin of the Navy". Asia-Pacific Defense Forum. Federation of American Scientists. Retrieved 2 February 2011.

- "Demon UAV jets into history by flying without flaps". Metro.co.uk. London: Associated Newspapers Limited. 28 September 2010.

Aircraft Manufacturing Process

The processes involved in the designing of aircrafts are termed as aircraft manufacturing process. It usually depends on factors such as customer demand, safety protocols and economic constraints. To provide a better understanding on the topic, a brief explanation on gull wing, aircraft flight control system and aeronautics is also given. This chapter is a compilation of the various branches of aircraft manufacturing that form an integral part of the broader subject matter.

Aircraft Design Process

The aircraft design process is the engineering design process by which aircraft are designed. These depend on many factors such as customer and manufacturer demand, safety protocols, physical and economic constraints etc. For some types of aircraft the design process is regulated by national airworthiness authorities. This article deals with powered aircraft such as airplanes and helicopter designs.

AST model in wind tunnel

Aircraft design is a compromise between many competing factors and constraints and accounts for existing designs and market requirements to produce the best aircraft.

Design Constraints

Purpose

The design process starts with the aircraft's intended purpose. Commercial airliners are designed for carrying a passenger or cargo payload, long range and greater fuel efficiency where as fighter jets are designed to perform high speed maneuvers and provide close support to ground troops. Some aircraft have specific missions, for instance, amphibious airplanes have a unique design that allows them to operate from both land and water, some fighters, like the Harrier Jump Jet, have VTOL (Vertical Take-off and Landing) ability, helicopters have the ability to hover over an area for a period of time.

The purpose may be to fit a specific requirement, e.g. as in the historical case of a British Air Ministry specification, or fill a perceived "gap in the market"; that is, a class or design of aircraft which does not yet exist, but for which there would be significant demand.

Aircraft Regulations

Another important factor that influences the design of the aircraft are the regulations put forth by national aviation airworthiness authorities.

Airports may also impose limits on aircraft, for instance, the maximum wingspan allowed for a conventional aircraft is 80 m to prevent collisions between aircraft while taxiing.

Financial Factors and Market

Budget limitations, market requirements and competition set constraints on the design process and comprise the non-technical influences on aircraft design along with environmental factors. Competition leads to companies striving for better efficiency in the design without compromising performance and incorporating new techniques and technology.

Environmental Factors

An increase in the number of aircraft also means greater carbon emissions. Environmental scientists have voiced concern over the main kinds of pollution associated with aircraft, mainly noise and emissions. Aircraft engines have been historically notorious for creating noise pollution and the expansion of airways over already congested and polluted cities have drawn heavy criticism, making it necessary to have environmental policies for aircraft noise. Noise also arises from the airframe, where the airflow directions are changed. Improved noise regulations have forced designers to create quieter engines and airframes. Emissions from aircraft include particulates, carbon dioxide (CO_2), Sulfur dioxide(SO_2), Carbon monoxide (CO), various oxides of nitrates and unburnt hydrocarbons. To combat the pollution, ICAO set recommendations in 1981 to control aircraft emissions. Newer, environmentally friendly fuels have been developed and the use of recyclable materials in manufacturing have helped reduce the ecological impact due to aircraft. Environmental limitations also affect airfield compatibility. Airports around the world have been built to suit the topography of the particular region. Space limitations, pavement design, runway end safety areas and the unique location of airport are some of the airport factors that influence aircraft design. However changes in aircraft design also influence airfield design as well, for instance, the recent introduction of new large aircraft (NLAs) such as the superjumbo Airbus A380, have led to airports worldwide redesigning their facilities to accommodate its large size and service requirements.

Safety

The high speeds, fuel tanks, atmospheric conditions at cruise altitudes, natural hazards (thunderstorms, hail and bird strikes) and human error are some of the many hazards that pose a threat to air travel.

Airworthiness is the standard by which aircraft are determined fit to fly. The responsibility for airworthiness lies with national aviation regulatory bodies, manufacturers, as well as owners and operators.

The International Civil Aviation Organization sets international standards and recommended practices for national authorities to base their regulations on The national regulatory authorities set standards for airworthiness, issue certificates to manufacturers and operators and the standards of personnel training. Every country has its own regulatory body such as the Federal Aviation Authority in USA, DGCA (Directorate General of Civil Aviation) in India, etc.

The aircraft manufacturer makes sure that the aircraft meets existing design standards, defines the operating limitations and maintenance schedules and provides support and maintenance throughout the operational life of the aircraft. The aviation operators include the passenger and cargo airliners, air forces and owners of private aircraft. They agree to comply with the regulations set by the regulatory bodies, understand the limitations of the aircraft as specified by the manufacturer, report defects and assist the manufacturers in keeping up the airworthiness standards.

Most of the design criticisms these days are built on crashworthiness. Even with the greatest attention to airworthiness, accidents still occur. Crashworthiness is the qualitative evaluation of how aircraft survive an accident. The main objective is to protect the passengers or valuable cargo from the damage caused by an accident. In the case of airliners the stressed skin of the pressurized fuselage provides this feature, but in the event of a nose or tail impact, large bending moments build all the way through the fuselage, causing fractures in the shell, causing the fuselage to break up into smaller sections. So the passenger aircraft are designed in such a way that seating arrangements are away from areas likely to be intruded in an accident, such as near a propeller, engine nacelle undercarriage etc. The interior of the cabin is also fitted with safety features such as oxygen masks that drop down in the event of loss of cabin pressure, lockable luggage compartments, safety belts, lifejackets, emergency doors and luminous floor strips. Aircraft are sometimes designed with emergency water landing in mind, for instance the Airbus A330 has a 'ditching' switch that closes valves and openings beneath the aircraft slowing the ingress of water.

Design Optimization

Aircraft designers normally rough-out the initial design with consideration of all the constraints on their design. Historically design teams used to be small, usually headed by a Chief Designer who knows all the design requirements and objectives and coordinated the team accordingly. As time progressed, the complexity of military and airline aircraft also grew. Modern military and airline design projects are of such a large scale that every design aspect is tackled by different teams and then brought together. In general aviation a large number of light aircraft are designed and built by amateur hobbyists and enthusiasts.

Computer-aided Design of Aircraft

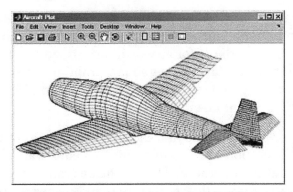

The external surfaces of an aircraft modelled in MATLAB

In the early years of aircraft design, designers generally used analytical theory to do the various engineering calculations that go into the design process along with a lot of experimentation. These calculations were labour-intensive and time-consuming. In the 1940s, several engineers started looking for ways to automate and simplify the calculation process and many relations and semi-empirical formulas were developed. Even after simplification, the calculations continued to be extensive. With the invention of the computer, engineers realized that a majority of the calculations could be automated, but the lack of design visualization and the huge amount of experimentation involved kept the field of aircraft design stagnant. With the rise of programming languages, engineers could now write programs that were tailored to design an aircraft. Originally this was done with mainframe computers and used low-level programming languages that required the user to be fluent in the language and know the architecture of the computer. With the introduction of personal computers, design programs began employing a more user-friendly approach.

Design Aspects

The main aspects of aircraft design are:

1. Aerodynamics

2. Propulsion

3. Controls

4. Mass

5. Structure

All aircraft designs involve compromises of these factors to achieve the design mission.

Wing Design

The wings of a fixed wing aircraft provide the necessary lift for take-off and cruise flight. Wing geometry affects every aspect of an aircraft's flight. The wing area will usually be dictated by aircraft performance requirements (e.g. field length) but the overall shape of the planform and other detail aspects may be influenced by wing layout factors. The wing can be mounted to the fuselage in high, low and middle positions. The wing design depends on many parameters such as selection of aspect ratio, taper ratio, sweepback angle, thickness ratio, section profile, washout and dihedral. The cross-sectional shape of the wing is its airfoil. The construction of the wing starts with the rib which defines the airfoil shape. Ribs can be made of wood, metal, plastic or even composites.

Fuselage

The fuselage is the part of the aircraft that contains the cockpit, passenger cabin or cargo hold.

Propulsion

Aircraft engine

Aircraft propulsion may be achieved by specially designed aircraft engines, adapted auto, motorcycle or snowmobile engines, electric engines or even human muscle power. The main parameters of engine design are:

- Maximum engine thrust available

- Fuel consumption

- Engine mass

- Engine geometry

The thrust provided by the engine must balance the drag at cruise speed and be greater than the drag to allow acceleration. The engine requirement varies with the type of aircraft. For instance, commercial airliners spend more time in cruise speed and need more engine efficiency. High-performance fighter jets need very high acceleration and therefore have very high thrust requirements.

Weight

The weight of the aircraft is the common factor that links all aspects of aircraft design such as aerodynamics, structure, and propulsion together. An aircraft's weight is derived from various factors such as empty weight, payload, useful load, etc. The various weights are used to then calculate the center of mass of the entire aircraft. The center of mass must fit within the established limits set by the manufacturer.

Structure

The aircraft structure focuses not only on strength, stiffness, durability (fatigue), fracture toughness, stability, but also on fail-safety, corrosion resistance, maintainability and ease of manufacturing. The structure must be able to withstand the stresses caused by cabin pressurization, if fitted, turbulence and engine or rotor vibrations.

Design Process and Simulation

The design of any aircraft starts out in three phases

Conceptual Design

The first design step, involves sketching a variety of possible aircraft configurations that meet the required design specifications. By drawing a set of configurations, designers seek to reach the design configuration that satisfactorily meets all requirements as well as go hand in hand with factors such as aerodynamics, propulsion, flight performance, structural and control systems. This is called design optimization. Fundamental aspects such as fuselage shape, wing configuration and location, engine size and type are all determined at this stage. Constraints to design like those mentioned above are all taken into account at this stage as well. The final product is a conceptual layout of the aircraft configuration on paper or computer screen, to be reviewed by engineers and other designers.

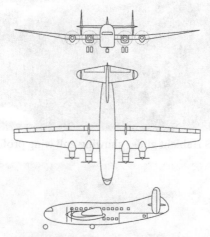

Conceptual design of a Breguet 673

Preliminary Design Phase

The design configuration arrived at in the conceptual design phase is then tweaked and remodeled to fit into the design parameters. In this phase, wind tunnel testing and computational fluid dynamic calculations of the flow field around the aircraft are done. Major structural and control analysis is also carried out in this phase. Aerodynamic flaws and structural instabilities if any are corrected and the final design is drawn and finalized. Then after the finalization of the design lies the key decision with the manufacturer or individual designing it whether to actually go ahead with the production of the aircraft. At this point several designs, though perfectly capable of flight and performance, might have been opted out of production due to their being economically nonviable.

Detail Design Phase

This phase simply deals with the fabrication aspect of the aircraft to be manufactured. It determines the number, design and location of ribs, spars, sections and other structural elements. All aerodynamic, structural, propulsion, control and performance aspects have already been covered in the preliminary design phase and only the manufacturing remains. Flight simulators for aircraft are also developed at this stage.

Gull Wing

The gull wing is an aircraft wing configuration with a prominent bend in the wing inner section towards the wing root. Its name is derived from the seabirds which it resembles. It has been incorporated in aircraft for many reasons.

DFS Habicht glider showing gull wing profile.

Laughing gull showing the wing shape emulated in gull wing aircraft.

Sailplanes

The gull wing was first seen on a glider when the Weltensegler flew in 1921. Its wings were externally braced and featured swept-back wingtips. After the aircraft broke up, killing its pilot, the design feature stayed out of popular use. The gull wing made a resurgence in 1930 with Alexander Lippisch's record-breaking *Fafnir*. Lippisch used the configuration for its increased wingtip clearance and the ill-founded belief it improved stability in turns. However, studies have shown that normal gull wing configurations have significantly less severe and more easily recoverable stalls. Inverted gull wings show the opposite stall behaviour, but both normal and inverted gull wings impede lift/drag ratio and climb performance. The true success of the Fafnir's gull wing lay primarily in its aesthetic value; the gull wing would be a staple of the high-performance sailplanes of the time, until the 1950s.

Notable Gull Wing Sailplanes:

- Bowlus Senior Albatross

- DFS Habicht

- DFS Kranich

- DFS Reiher

- Göppingen Gö 3 *Minimoa*

- Lawrence Tech IV "Yankee Doodle"

- Lippisch Fafnir

- Ross RS-1 Zanonia

- Schweyer Rhönsperber

- Slingsby Kite

- Weltensegler

Seaplanes

Beriev Be-12 seaplane with gull wing profile

The gull wing design found its way into seaplanes by the early 1930s. As engine power increased, so did the need for large propellers that could effectively convert power to thrust. The gull wing allowed designers to ensure adequate propeller tip clearance over the water by placing the engines on the highest point of the wing. The alternative was placing the engine on a pylon. Possibly the first flying boat to utilize the gull wing configuration was the Short Knuckleduster, which flew in 1933. The Dornier Do 26, a high-speed airliner and transport platform, of which 6 aircraft were built, flew in 1938. The configuration was also used on the US Navy's PBM Mariner and P5M Marlin maritime patrol aircraft. The emergence of long range, land-based jets in the 1950s and the subsequent demise of the seaplane prevented widespread use of the gull wing, although it was still used in some post-war designs, like Beriev Be-12 *Chaika* (the name means 'the gull' in Russian).

Examples:

- Beriev Be-6

- Dornier Do 26

- Martin P5M Marlin

- Piaggio P.136

- Short Knuckleduster

Landplanes

The gull wing design found its way into landplanes in the late 1920s, with Polish inventor Zygmunt Pulawski designing the PZL P.1 in 1928. The arrangement he devised is occasionally known as the "Pulawski Wing" or the "Polish wing". The gull wing was used to improve visibility in a high wing arrangement, because such wing could be thinnest by the fuselage, and in theory should limit pilot's view no more than A-pillars of a windscreen in a car body. It was used in fighter aircraft like PZL P.11 and Polikarpov I-15.

Examples:

- PZL P.11

- Polikarpov I-153

Inverted Gull Wing

Junkers Ju 87 Stuka German ground-attack aircraft of WWII

F4U Corsair landing on USS *Bunker Hill*

Aichi B7A carrying torpedo.

The inverted gull wing was also developed in the 1930s and was chiefly used on single engine military aircraft with increasing powerful engines. Before contra-rotating propellers came into use, such powers required larger diameter propellers but clearance between the propeller tip and ground had to be maintained. Long landing gear legs are heavy, bulky, and weaker than their shorter counterparts. The Vought F4U Corsair, designed from the onset as a carrier-based fighter, not only had the largest propeller of any U.S. fighter, but was also expected to face rough landings aboard a pitching carrier deck. The inverted gull wing allowed the landing gear to be short and strong, and to retract straight back, improving internal wing space. An additional aerodynamic advantage was that the wing/fuselage connection is perpendicular and has inherently lower drag than any other connection.

Another reason for having an inverted gull wing is to permit clearance for a large external bomb load, as on the Junkers Ju-87 Stuka.

Aircraft Flight Control System

A conventional fixed-wing aircraft flight control system consists of flight control surfaces, the respective cockpit controls, connecting linkages, and the necessary operating mechanisms to control an aircraft's direction in flight. Aircraft engine controls are also considered as flight controls as they change speed.

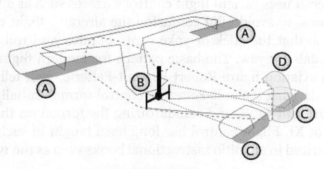

A typical aircraft's primary flight controls in motion

The fundamentals of aircraft controls are explained in flight dynamics. This article centers on the operating mechanisms of the flight controls. The basic system in use on aircraft first appeared in a readily recognizable form as early as April 1908, on Louis Blériot's Blériot VIII pioneer-era monoplane design.

Cockpit Controls

Primary Controls

Generally, the primary cockpit flight controls are arranged as follows:

- a control yoke (also known as a control column), centre stick or side-stick (the latter two also colloquially known as a control or joystick), governs the aircraft's roll and pitch by moving the ailerons (or activating wing warping on some very early aircraft designs) when turned or deflected left and right, and moves the elevators when moved backwards or forwards

- rudder pedals, or the earlier, pre-1919 "rudder bar", to control yaw, which move the rudder; left foot forward will move the rudder left for instance.

- throttle controls to control engine speed or thrust for powered aircraft.

The control yokes also vary greatly amongst aircraft. There are yokes where roll is controlled by rotating the yoke clockwise/counterclockwise (like steering a car) and pitch is controlled by tilting the control column towards you or away from you, but in others the pitch is controlled by sliding the yoke into and out of the instrument panel (like most Cessnas, such as the 152 and 172), and in some the roll is controlled by sliding the whole yoke to the left and right (like the Cessna 162). Centre sticks also vary between aircraft. Some are directly connected to the control surfaces using cables, others (fly-by-wire airplanes) have a computer in between which then controls the electrical actuators.

Blériot VIII at Issy-les-Moulineaux, the first flightworthy aircraft design to have the initial form of modern flight controls for the pilot

Even when an aircraft uses variant flight control surfaces such as a V-tail ruddervator, flaperons, or elevons, to avoid pilot confusion the aircraft's flight control system will still be designed so that the stick or yoke controls pitch and roll conventionally, as will the rudder pedals for yaw. The basic pattern for modern flight controls was pioneered by French aviation figure Robert Esnault-Pelterie, with fellow French aviator Louis Blériot popularizing Esnault-Pelterie's control format initially on Louis' Blériot VIII monoplane in April 1908, and standardizing the format on the July 1909 Channel-crossing Blériot XI. Flight control has long been taught in such fashion for many decades, as popularized in ab initio instructional books such as the 1944 work Stick and Rudder.

In some aircraft, the control surfaces are not manipulated with a linkage. In ultralight aircraft and motorized hang gliders, for example, there is no mechanism at all. Instead, the pilot just grabs the lifting surface by hand (using a rigid frame that hangs from its underside) and moves it.

Secondary Controls

In addition to the primary flight controls for roll, pitch, and yaw, there are often secondary controls available to give the pilot finer control over flight or to ease the workload. The most commonly available control is a wheel or other device to control elevator trim, so that the pilot does not have to maintain constant backward or forward pressure to hold a specific pitch attitude (other types of trim, for rudder and ailerons, are common on larger aircraft but may also appear on smaller ones). Many aircraft have wing flaps, controlled by a switch or a mechanical lever or in some cases are fully automatic by computer control, which alter the shape of the wing for improved control at the slower speeds used for take-off and landing. Other secondary flight control systems may be available, including slats, spoilers, air brakes and variable-sweep wings.

Flight Control Systems

Mechanical

de Havilland Tiger Moth elevator and rudder cables

Mechanical or manually operated flight control systems are the most basic method of controlling an aircraft. They were used in early aircraft and are currently used in small aircraft where the aerodynamic forces are not excessive. Very early aircraft, such as the Wright Flyer I, Blériot XI and Fokker Eindecker used a system of wing warping where no conventionally hinged control surfaces were used on the wing, and sometimes not even for pitch control as on the Wright Flyer I and original versions of the 1909 Etrich Taube, which only had a hinged/pivoting rudder in addition to the warping-operated pitch and roll controls. A manual flight control system uses a collection of mechanical parts such as pushrods, tension cables, pulleys, counterweights, and sometimes chains to transmit the forces applied to the cockpit controls directly to the control surfaces. Turnbuckles are often used to adjust control cable tension. The Cessna Skyhawk is a typical example of an aircraft that uses this type of system. Gust locks are often used on parked aircraft with mechanical systems to protect the control surfaces and linkages from damage from wind. Some aircraft have gust locks fitted as part of the control system.

Increases in the control surface area required by large aircraft or higher loads caused by high airspeeds in small aircraft lead to a large increase in the forces needed to move them, consequently complicated mechanical gearing arrangements were developed to extract maximum mechanical advantage in order to reduce the forces required from the pilots. This arrangement can be found on bigger or higher performance propeller aircraft such as the Fokker 50.

Some mechanical flight control systems use servo tabs that provide aerodynamic assistance. Servo tabs are small surfaces hinged to the control surfaces. The flight control mechanisms move these tabs, aerodynamic forces in turn move, or assist the movement of the control surfaces reducing the amount of mechanical forces needed. This arrangement was used in early piston-engined transport aircraft and in early jet transports. The Boeing 737 incorporates a system, whereby in the unlikely event of total hydraulic system failure, it automatically and seamlessly reverts to being controlled via servo-tab.

Hydro-mechanical

The complexity and weight of mechanical flight control systems increase considerably with the size and performance of the aircraft. Hydraulically powered control surfaces help to overcome these limitations. With hydraulic flight control systems, the aircraft's size and performance are limited by economics rather than a pilot's muscular strength. At first, only-partially boosted systems were used in which the pilot could still feel some of the aerodynamic loads on the control surfaces (feedback).

A hydro-mechanical flight control system has two parts:

- The *mechanical circuit*, which links the cockpit controls with the hydraulic circuits. Like the mechanical flight control system, it consists of rods, cables, pulleys, and sometimes chains.

- The *hydraulic circuit*, which has hydraulic pumps, reservoirs, filters, pipes, valves and actuators. The actuators are powered by the hydraulic pressure generated by the pumps in the hydraulic circuit. The actuators convert hydraulic pressure into control surface movements. The electro-hydraulic servo valves control the movement of the actuators.

The pilot's movement of a control causes the mechanical circuit to open the matching servo valve in the hydraulic circuit. The hydraulic circuit powers the actuators which then move the control surfaces. As the actuator moves, the servo valve is closed by a mechanical feedback linkage - one that stops movement of the control surface at the desired position.

This arrangement was found in the older-designed jet transports and in some high-performance aircraft. Examples include the Antonov An-225 and the Lockheed SR-71.

Artificial Feel Devices

With purely mechanical flight control systems, the aerodynamic forces on the control surfaces are transmitted through the mechanisms and are felt directly by the pilot, allowing tactile feedback of airspeed. With hydromechanical flight control systems, however, the load on the surfaces cannot be felt and there is a risk of overstressing the aircraft through excessive control surface movement. To overcome this problem, artificial feel systems can be used. For example, for the controls of the RAF's Avro Vulcan jet bomber and the RCAF's Avro Canada CF-105 Arrow supersonic interceptor (both 1950s-era designs), the required force feedback was achieved by a spring device. The fulcrum of this device was moved in proportion to the square of the air speed (for the elevators) to give increased resistance at higher speeds. For the controls of the American Vought F-8 Crusader and the LTV A-7 Corsair II warplanes, a 'bob-weight' was used in the pitch axis of the control stick, giving force feedback that was proportional to the airplane's normal acceleration.

Stick Shaker

A stick shaker is a device (available in some hydraulic aircraft) that is attached to the control column, which shakes the control column when the aircraft is about to stall. Also in some aircraft like the McDonnell Douglas DC-10 there is/was a back-up electrical power supply that the pilot can turn on to re-activate the stick shaker in case the hydraulic connection to the stick shaker is lost.

Fly-by-wire Control Systems

A fly-by-wire (FBW) system replaces manual flight control of an aircraft with an electronic interface. The movements of flight controls are converted to electronic signals transmitted by wires (hence the fly-by-wire term), and flight control computers determine how to move the actuators at each control surface to provide the expected response. Commands from the computers are also input without the pilot's knowledge to stabilize the aircraft and perform other tasks. Electronics for aircraft flight control systems are part of the field known as avionics.

Fly-by-optics, also known as *fly-by-light*, is a further development using fiber optic cables. This has an added advantage when sensitive electro-magnetic sensors will be operating aboard the aircraft.

Research

Several technology research and development efforts exist to integrate the functions of flight control systems such as ailerons, elevators, elevons, flaps, and flaperons into wings to perform the aerodynamic purpose with the advantages of less: mass, cost, drag, inertia (for faster, stronger control response), complexity (mechanically simpler, fewer moving parts or surfaces, less maintenance), and radar cross section for stealth. These may be used in many unmanned aerial vehicles (UAVs) and 6th generation fighter aircraft. Two promising approaches are flexible wings, and fluidics.

Flexible Wings

In flexible wings, much or all of a wing surface can change shape in flight to deflect air flow much like an ornithopter. Adaptive compliant wings are a military and commercial effort. The X-53 Active Aeroelastic Wing was a US Air Force, NASA, and Boeing effort.

Fluidics

In fluidics, forces in vehicles occur via circulation control, in which larger more complex mechanical parts are replaced by smaller simpler fluidic systems (slots which emit air flows) where larger forces in fluids are diverted by smaller jets or flows of fluid intermittently, to change the direction of vehicles. In this use, fluidics promises lower mass, costs (up to 50% less), and very low inertia and response times, and simplicity. This was demonstrated in the Demon UAV, which flew for the first time, in the UK, in September 2010.

Aviation

Aviation is the practical aspect or art of aeronautics, being the design, development, production, operation and use of aircraft, especially heavier than air aircraft. The word *aviation* was coined by French writer and former naval officer Gabriel La Landelle in 1863, from the verb *avier* (synonymous flying), itself derived from the Latin word *avis* ("bird") and the suffix *-ation*.

Early Beginnings

There are early legends of human flight such as the story of Icarus in Greek myth and Jamshid in Persian myth, and later, somewhat more credible claims of short-distance human flights appear, such as the flying automaton of Archytas of Tarentum (428–347 BC), the winged flights of Abbas Ibn Firnas (810–887), Eilmer of Malmesbury (11th century), and the hot-air Passarola of Bartholomeu Lourenço de Gusmão (1685–1724).

Lighter than Air

LZ 129 Hindenburg at Lakehurst Naval Air Station, 1936

The modern age of aviation began with the first untethered human lighter-than-air flight on November 21, 1783, of a hot air balloon designed by the Montgolfier brothers. The practicality of balloons was limited because they could only travel downwind. It was immediately recognized that a steerable, or dirigible, balloon was required. Jean-Pierre Blanchard flew the first human-powered dirigible in 1784 and crossed the English Channel in one in 1785.

Rigid airships became the first aircraft to transport passengers and cargo over great distances. The best known aircraft of this type were manufactured by the German Zeppelin company.

The most successful Zeppelin was the Graf Zeppelin. It flew over one million miles, including an around-the-world flight in August 1929. However, the dominance of the Zeppelins over the airplanes of that period, which had a range of only a few hundred miles, was diminishing as airplane design advanced. The "Golden Age" of the airships ended on May 6, 1937 when the Hindenburg caught fire, killing 36 people. The cause of the Hindenburg accident was initially blamed on the use of hydrogen instead of helium as the lift gas. An internal investigation by the manufacturer revealed the coating used to protect the covering material over the frame was highly flammable and allowed static electricity to build up in the airship. Changes to the coating formulation reduced the risk of further Hindenburg type accidents. Although there have been periodic initiatives to revive their use, airships have seen only niche application since that time.

Heavier than Air

In 1799 Sir George Cayley set forth the concept of the modern airplane as a fixed-wing flying machine with separate systems for lift, propulsion, and control. Early dirigible developments included machine-powered propulsion (Henri Giffard, 1852), rigid frames (David Schwarz, 1896) and improved speed and maneuverability (Alberto Santos-Dumont, 1901)

First flight by the Wright Brothers, December 17, 1903

There are many competing claims for the earliest powered, heavier-than-air flight. The first recorded powered flight was carried out by Clément Ader on October 9,

1890 when he reportedly made the first manned, powered, heavier-than-air flight of a significant distance (50 m (160 ft)) but insignificant altitude from level ground in his bat-winged, fully self-propelled fixed-wing aircraft, the Ader Éole. Seven years later, on 14 October 1897, Ader's Avion III was tested without success in front of two officials from the French War ministry. The report on the trials was not publicized until 1910, as they had been a military secret. In November 1906 Ader claimed to have made a successful flight on 14 October 1897, achieving an "uninterrupted flight" of around 300 metres (980 feet) on. Although widely believed at the time, these claims were later discredited.

However, the most widely accepted date is December 17, 1903 by the Wright brothers. The Wright brothers were the first to fly in a powered and controlled aircraft. Previous flights were gliders (control but no power) or free flight (power but no control), but the Wright brothers combined both, setting the new standard in aviation records. Following this, the widespread adoption of ailerons rather than wing warping made aircraft much easier to control, and only a decade later, at the start of World War I, heavier-than-air powered aircraft had become practical for reconnaissance, artillery spotting, and even attacks against ground positions.

Aircraft began to transport people and cargo as designs grew larger and more reliable. The Wright brothers took aloft the first passenger, Charles Furnas, one of their mechanics, on May 14, 1908.

During the 1920s and 1930s great progress was made in the field of aviation, including the first transatlantic flight of Alcock and Brown in 1919, Charles Lindbergh's solo transatlantic flight in 1927, and Charles Kingsford Smith's transpacific flight the following year. One of the most successful designs of this period was the Douglas DC-3, which became the first airliner to be profitable carrying passengers exclusively, starting the modern era of passenger airline service. By the beginning of World War II, many towns and cities had built airports, and there were numerous qualified pilots available. The war brought many innovations to aviation, including the first jet aircraft and the first liquid-fueled rockets.

NASA's Helios researches solar powered flight.

After World War II, especially in North America, there was a boom in general aviation, both private and commercial, as thousands of pilots were released from military service and many inexpensive war-surplus transport and training aircraft became available. Manufacturers such as Cessna, Piper, and Beechcraft expanded production to provide light aircraft for the new middle-class market.

By the 1950s, the development of civil jets grew, beginning with the de Havilland Comet, though the first widely used passenger jet was the Boeing 707, because it was much more economical than other aircraft at that time. At the same time, turboprop propulsion began to appear for smaller commuter planes, making it possible to serve small-volume routes in a much wider range of weather conditions.

Since the 1960s composite material airframes and quieter, more efficient engines have become available, and Concorde provided supersonic passenger service for more than two decades, but the most important lasting innovations have taken place in instrumentation and control. The arrival of solid-state electronics, the Global Positioning System, satellite communications, and increasingly small and powerful computers and LED displays, have dramatically changed the cockpits of airliners and, increasingly, of smaller aircraft as well. Pilots can navigate much more accurately and view terrain, obstructions, and other nearby aircraft on a map or through synthetic vision, even at night or in low visibility.

On June 21, 2004, SpaceShipOne became the first privately funded aircraft to make a spaceflight, opening the possibility of an aviation market capable of leaving the Earth's atmosphere. Meanwhile, flying prototypes of aircraft powered by alternative fuels, such as ethanol, electricity, and even solar energy, are becoming more common.

Operations of Aircraft

Civil Aviation

Diamond Aircraft Industries Diamond D-Jet (very light jet)

Civil aviation includes all non-military flying, both general aviation and scheduled air transport.

Air Transport

Northwest Airlines Airbus A330-323X

There are five major manufacturers of civil transport aircraft (in alphabetical order):

- Airbus, based in Europe
- Boeing, based in the United States
- Bombardier, based in Canada
- Embraer, based in Brazil
- United Aircraft Corporation, based in Russia

Boeing, Airbus, Ilyushin and Tupolev concentrate on wide-body and narrow-body jet airliners, while Bombardier, Embraer and Sukhoi concentrate on regional airliners. Large networks of specialized parts suppliers from around the world support these manufacturers, who sometimes provide only the initial design and final assembly in their own plants. The Chinese ACAC consortium will also soon enter the civil transport market with its Comac ARJ21 regional jet.

Until the 1970s, most major airlines were flag carriers, sponsored by their governments and heavily protected from competition. Since then, open skies agreements have resulted in increased competition and choice for consumers, coupled with falling prices for airlines. The combination of high fuel prices, low fares, high salaries, and crises such as the September 11, 2001 attacks and the SARS epidemic have driven many older airlines to government-bailouts, bankruptcy or mergers. At the same time, low-cost carriers such as Ryanair, Southwest and Westjet have flourished.

General Aviation

General aviation includes all non-scheduled civil flying, both private and commercial. General aviation may include business flights, air charter, private aviation, flight training, ballooning, parachuting, gliding, hang gliding, aerial photography, foot-launched powered hang gliders, air ambulance, crop dusting, charter flights, traffic reporting, police air patrols and forest fire fighting.

1947 Cessna 120

A weight-shift ultralight aircraft, the Air Creation Tanarg

Each country regulates aviation differently, but general aviation usually falls under different regulations depending on whether it is private or commercial and on the type of equipment involved.

Many small aircraft manufacturers serve the general aviation market, with a focus on private aviation and flight training.

The most important recent developments for small aircraft (which form the bulk of the GA fleet) have been the introduction of advanced avionics (including GPS) that were formerly found only in large airliners, and the introduction of composite materials to make small aircraft lighter and faster. Ultralight and homebuilt aircraft have also become increasingly popular for recreational use, since in most countries that allow private aviation, they are much less expensive and less heavily regulated than certified aircraft.

The largest aircraft to be built, to date, is the Antonov An-225. This aircraft comes from the Ukraine, and it was built back in the 1980s. This aircraft includes 6 engines, mounted on the wing. Its wingspan is 88 metres (289 feet) and it is 84 metres (276 feet) long. This aircraft holds the world payload record, after it transported 428,834 pounds worth of goods. Weighing in at 1.4 million pounds, it is also the heaviest aircraft to be built.

Military Aviation

Simple balloons were used as surveillance aircraft as early as the 18th century. Over the years, military aircraft have been built to meet ever increasing capability requirements.

Manufacturers of military aircraft compete for contracts to supply their government's arsenal. Aircraft are selected based on factors like cost, performance, and the speed of production.

The Lockheed SR-71 remains unsurpassed in many areas of performance.

Types of Military Aviation

- Fighter aircraft's primary function is to destroy other aircraft. (e.g. Sopwith Camel, A6M Zero, F-15, MiG-29, Su-27, and F-22).

- Ground attack aircraft are used against tactical earth-bound targets. (e.g. Junkers Stuka, A-10, Il-2, J-22 Orao, AH-64 and Su-25).

- Bombers are generally used against more strategic targets, such as factories and oil fields. (e.g. Zeppelin, Tu-95, Mirage IV, and B-52).

- Transport aircraft are used to transport hardware and personnel. (e.g. C-17 Globemaster III, C-130 Hercules and Mil Mi-26).

- Surveillance and reconnaissance aircraft obtain information about enemy forces. (e.g. Rumpler Taube, Mosquito, U-2, OH-58 and MiG-25R).

- Unmanned aerial vehicles (UAVs) are used primarily as reconnaissance fixed-wing aircraft, though many also carry payloads. Cargo aircraft are in development. (e.g. RQ-7B Shadow, MQ-8 Fire Scout, and MQ-1C Gray Eagle).

- Missiles deliver warheads, normally explosives, but also things like leaflets.

Air Safety

Aviation Accidents and Incidents

An *aviation accident* is defined by the Convention on International Civil Aviation Annex 13 as an occurrence associated with the operation of an aircraft which takes place between the time any person boards the aircraft with the intention of flight until such time as all such persons have disembarked, in which a person is fatally or seriously injured, the aircraft sustains damage or structural failure or the aircraft is missing or is completely inaccessible.

A USAF Thunderbird pilot ejecting from his F-16 aircraft at an airshow in 2003

The first fatal aviation accident occurred in a Wright Model A aircraft at Fort Myer, Virginia, USA, on September 17, 1908, resulting in injury to the pilot, Orville Wright, and death of the passenger, Signal Corps Lieutenant Thomas Selfridge.

An *aviation incident* is defined as an occurrence, other than an accident, associated with the operation of an aircraft that affects or could affect the safety of operations.

An accident in which the damage to the aircraft is such that it must be written off, or in which the plane is destroyed, is called a *hull loss accident*.

Air Traffic Control

Air traffic control towers at Amsterdam Airport

Air traffic control (ATC) involves communication with aircraft to help maintain separation — that is, they ensure that aircraft are sufficiently far enough apart horizontally or vertically for no risk of collision. Controllers may co-ordinate position reports provided by pilots, or in high traffic areas (such as the United States) they may use radar to see aircraft positions.

There are generally four different types of ATC:

- center controllers, who control aircraft en route between airports

- control towers (including tower, ground control, clearance delivery, and other services), which control aircraft within a small distance (typically 10–15 km horizontal, and 1,000 m vertical) of an airport.

- oceanic controllers, who control aircraft over international waters between continents, generally without radar service.

- terminal controllers, who control aircraft in a wider area (typically 50–80 km) around busy airports.

ATC is especially important for aircraft flying under instrument flight rules (IFR), where they may be in weather conditions that do not allow the pilots to see other aircraft. However, in very high-traffic areas, especially near major airports, aircraft flying under visual flight rules (VFR) are also required to follow instructions from ATC.

In addition to separation from other aircraft, ATC may provide weather advisories, terrain separation, navigation assistance, and other services to pilots, depending on their workload.

ATC do not control all flights. The majority of VFR flights in North America are not required to talk to ATC (unless they are passing through a busy terminal area or using a major airport), and in many areas, such as northern Canada and low altitude in northern Scotland, Air traffic control services are not available even for IFR flights at lower altitudes.

Environmental Impact

Water vapor contrails left by high-altitude jet airliners. These may contribute to cirrus cloud formation.

Like all activities involving combustion, operating powered aircraft (from airliners to hot air balloons) release soot and other pollutants into the atmosphere. Greenhouse gases such as carbon dioxide (CO_2) are also produced. In addition, there are environmental impacts specific to aviation:

- Aircraft operating at high altitudes near the tropopause (mainly large jet airliners) emit aerosols and leave contrails, both of which can increase cirrus cloud formation — cloud cover may have increased by up to 0.2% since the birth of aviation.

- Aircraft operating at high altitudes near the tropopause can also release chemicals that interact with greenhouse gases at those altitudes, particularly nitrogen compounds, which interact with ozone, increasing ozone concentrations.

- Most light piston aircraft burn avgas, which contains tetraethyllead (TEL). Some lower-compression piston engines can operate on unleaded mogas, and turbine engines and diesel engines — neither of which requires lead — are appearing on some newer light aircraft.

Another environmental impact of aviation is noise pollution, mainly caused by aircraft taking off and landing.

Aeronautics

Space Shuttle *Atlantis* on a Shuttle Carrier Aircraft.

Lighter than air geostationary airship telecommunications satellite

Aeronautics (from the ancient Greek words ⬜ὴρ *āēr*, which means "air", and ναυτικὴ *nautikē* which means "navigation", i.e. "navigation into the air") is the science or art involved with the study, design, and manufacturing of air flight capable machines, and the techniques of operating aircraft and rockets within the atmosphere. The British Royal Aeronautical Society identifies the aspects of "aeronautical Art, Science and Engineering" and "the profession of Aeronautics (which expression includes Astronautics)."

While the term—literally meaning "sailing the air"—originally referred solely to the science of *operating* the aircraft, it has since been expanded to include technology, business and other aspects related to aircraft. The term "aviation" is sometimes used interchangeably with aeronautics, although "aeronautics" includes lighter-than-air craft such as airships, and includes ballistic vehicles while "aviation" technically does not.

A significant part of aeronautical science is a branch of dynamics called aerodynamics, which deals with the motion of air and the way that it interacts with objects in motion, such as an aircraft.

History

Early Ideas

Attempts to fly without any real aeronautical understanding have been made from the earliest times, typically by constructing wings and jumping from a tower with crippling or lethal results.

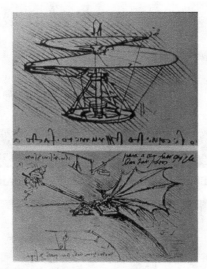

Designs for flying machines by Leonardo da Vinci, circa 1490

Wiser investigators sought to gain some rational understanding through the study of bird flight. An early example appears in ancient Egyptian texts. Later medieval Islamic scientists also made such studies. The founders of modern aeronautics, Leonardo da Vinci in the Renaissance and Cayley in 1799, both began their investigations with studies of bird flight.

Man-carrying kites are believed to have been used extensively in ancient China. In 1282 the European explorer Marco Polo described the Chinese techniques then current. The Chinese also constructed small hot air balloons, or lanterns, and rotary-wing toys.

An early European to provide any scientific discussion of flight was Roger Bacon, who described principles of operation for the lighter-than-air balloon and the flapping-wing ornithopter, which he envisaged would be constructed in the future. The lifting medium for his balloon would be an "aether" whose composition he did not know.

In the late fifteenth century, Leonardo da Vinci followed up his study of birds with designs for some of the earliest flying machines, including the flapping-wing ornithopter and the rotating-wing helicopter. Although his designs were rational, they were not based on particularly good science. Many of his designs, such as a four-person screw-type helicopter, have severe flaws. He did at least understand that "An object offers as much resistance to the air as the air does to the object." (Newton would not publish the Third law of motion until 1687.) His analysis led to the realisation that manpower alone was not sufficient for sustained flight, and his later designs included a mechanical power source such as a spring. Da Vinci's work was lost after his death and did not reappear until it had been overtaken by the work of George Cayley.

Balloon Flight

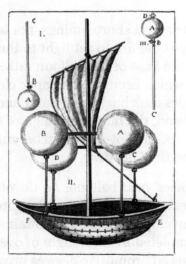

Francesco Lana de Terzi's flying boat concept c.1670

The modern era of lighter-than-air flight began early in the 17th century with Galileo's experiments in which he showed that air has weight. Around 1650 Cyrano de Bergerac wrote some fantasy novels in which he described the principle of ascent using a substance (dew) he supposed to be lighter than air, and descending by releasing a controlled amount of the substance. Francesco Lana de Terzi measured the pressure of air at sea level and in 1670 proposed the first scientifically credible lifting medium in the form of hollow metal spheres from which all the air had been pumped out. These would be lighter than the displaced air and able to lift an airship. His proposed methods of controlling height are still in use today; by carrying ballast which may be dropped overboard to gain height, and by venting the lifting containers to lose height. In practice de Terzi's spheres would have collapsed under air pressure, and further developments had to wait for more practicable lifting gases.

From the mid-18th century the Montgolfier brothers in France began experimenting with balloons. Their balloons were made of paper, and early experiments using steam as the lifting gas were short-lived due to its effect on the paper as it condensed. Mistaking smoke for a kind of steam, they began filling their balloons with hot smoky air which they called "electric smoke" and, despite not fully understanding the principles at work, made some successful launches and in 1783 were invited to give a demonstration to the French Academie des Sciences.

Meanwhile, the discovery of hydrogen led Joseph Black in c. 1780 to propose its use as a lifting gas, though practical demonstration awaited a gastight balloon material. On hearing of the Montgolfier Brothers' invitation, the French Academy member Jacques Charles offered a similar demonstration of a hydrogen balloon. Charles and two craftsmen, the Robert brothers, developed a gastight material of rubberised silk for the envelope. The hydrogen gas was to be generated by chemical reaction during the filling process.

The Montgolfier designs had several shortcomings, not least the need for dry weather and a tendency for sparks from the fire to set light to the paper balloon. The manned design had a gallery around the base of the balloon rather than the hanging basket of the first, unmanned design, which brought the paper closer to the fire. On their free flight, De Rozier and d'Arlandes took buckets of water and sponges to douse these fires as they arose. On the other hand, the manned design of Charles was essentially modern. As a result of these exploits, the hot-air ballon became known as the *Montgolfière* type and the hydrogen balloon the *Charlière*.

Charles and the Robert brothers' next balloon, *La Caroline*, was a Charlière that followed Jean Baptiste Meusnier's proposals for an elongated dirigible balloon, and was notable for having an outer envelope with the gas contained in a second, inner ballonet. On 19 September 1784, it completed the first flight of over 100 km, between Paris and Beuvry, despite the man-powered propulsive devices proving useless.

In an attempt the next year to provide both endurance and controllability, de Rozier developed a balloon having both hot air and hydrogen gas bags, a design which was soon named after him as the *Rozière*. The principle was to use the hydrogen section for constant lift and to navigate vertically by heating and allowing to cool the hot air section, in order to catch the most favourable wind at whatever altitude it was blowing. The balloon envelope was made of goldbeaters skin. The first flight ended in disaster and the approach has seldom been used since.

Cayley and the Foundation of Modern Aeronautics

Sir George Cayley (1773-1857) is widely acknowledged as the founder of modern aeronautics. He was first called the "father of the aeroplane" in 1846 and Henson called him the "father of aerial navigation." He was the first true scientific aerial investigator to publish his work, which included for the first time the underlying principles and forces of flight.

In 1809 he began the publication of a landmark three-part treatise titled "On Aerial Navigation" (1809–1810). In it he wrote the first scientific statement of the problem, "The whole problem is confined within these limits, viz. to make a surface support a given weight by the application of power to the resistance of air." He identified the four vector forces that influence an aircraft: *thrust*, *lift*, *drag* and *weight* and distinguished stability and control in his designs.

He developed the modern conventional form of the fixed-wing aeroplane having a stabilising tail with both horizontal and vertical surfaces, flying gliders both unmanned and manned.

He introduced the use of the whirling arm test rig to investigate the aerodynamics of flight, using it to discover the benefits of the curved or cambered aerofoil over the flat wing he had used for his first glider. He also identified and described the importance of dihedral, diagonal bracing and drag reduction, and contributed to the understanding and design of ornithopters and parachutes.

Another significant invention was the tension-spoked wheel, which he devised in order to create a light, strong wheel for aircraft undercarriage.

The 19th Century

During the 19th century Cayley's ideas were refined, proved and expanded on. Important investigators included Otto Lilienthal and Horatio Phillips.

Branches

The Eurofighter Typhoon.

Antonov An-225 *Mriya*, the largest aeroplane ever built.

Aeronautics may be divided into three main branches comprising Aviation, Aeronautical science and Aeronautical engineering.

Aviation

Aviation is the art or practice of aeronautics. Historically aviation meant only heavier-than-air flight, but nowadays it includes flying in balloons and airships.

Aeronautical Science

Aeronautical science covers the practical theory of aeronautics and aviation, including operations, navigation, air safety and human factors.

A candidate pilot is likely to study for a qualification in aeronautical science.

Aeronautical Engineering

Aeronautical engineering covers the design and construction of aircraft, including how they are powered, how they are used and how they are controlled for safe operation.

A major part of aeronautical engineering is aerodynamics, the science of passing through the air.

With the increasing activity in spaceflight, nowadays aeronautics and astronautics are often combined as aerospace engineering.

Aerodynamics

The science of aerodynamics deals with the motion of air and the way that it interacts with objects in motion, such as an aircraft.

The study of aerodynamics falls broadly into three areas:

Incompressible flow occurs where the air simply moves to avoid objects, typically at subsonic speeds below that of sound (Mach 1).

Compressible flow occurs where shock waves appear at points where the air becomes compressed, typically at speeds above Mach 1.

Transonic flow occurs in the intermediate speed range around Mach 1, where the airflow over an object may be locally subsonic at one point and locally supersonic at another.

Rocketry

A rocket or rocket vehicle is a missile, spacecraft, aircraft or other vehicle which obtains thrust from a rocket engine. In all rockets, the exhaust is formed entirely from propellants carried within the rocket before use. Rocket engines work by action and reaction. Rocket engines push rockets forwards simply by throwing their exhaust backwards extremely fast.

Rockets for military and recreational uses date back to at least 13th century China. Significant scientific, interplanetary and industrial use did not occur until the 20th century, when rocketry was the enabling technology of the Space Age, including setting foot on the moon.

Rockets are used for fireworks, weaponry, ejection seats, launch vehicles for artificial satellites, human spaceflight and exploration of other planets. While comparatively inefficient for low speed use, they are very lightweight and powerful, capable of generating large accelerations and of attaining extremely high speeds with reasonable efficiency.

Chemical rockets are the most common type of rocket and they typically create their exhaust by the combustion of rocket propellant. Chemical rockets store a large amount of energy in an easily released form, and can be very dangerous. However, careful design, testing, construction and use minimizes risks.

References

- Sutton, George (2001). "1". Rocket Propulsion Elements (7th ed.). Chichester: John Wiley & Sons. ISBN 978-0-471-32642-7.
- Cayley, George. "On Aerial Navigation" Part 1, Part 2, Part 3 Nicholson's Journal of Natural Philosophy, 1809–1810. (Via NASA). Raw text. Retrieved: 30 May 2010.

Flight Test: An Integrated Study

Data is gathered during the flight of an aircraft. The collection of the data and the analysis of it is known as flight test. Cooper-Harper rating scale, flying qualities, drop test and airframe are some of the topics discussed in the following content. The aspects elucidated in this chapter are of vital importance, and provide a better understanding of aerospace engineering.

Flight Test

Flight testing is a branch of aeronautical engineering that develops and gathers data during flight of an aircraft, or atmospheric testing of launch vehicles and reusable spacecraft, and then analyzes the data to evaluate the aerodynamic flight characteristics of the vehicle in order to validate the design, including safety aspects.

The flight test phase accomplishes two major tasks: 1) finding and fixing any design problems and then 2) verifying and documenting the vehicle capabilities for government certification or customer acceptance. The flight test phase can range from the test of a single new system for an existing vehicle to the complete development and certification of a new aircraft, launch vehicle, or reusable spacecraft. Therefore, the duration of a particular flight test program can vary from a few weeks to many years.

Aircraft Flight Test

Civil Aircraft

There are typically two categories of flight test programs – commercial and military. Commercial flight testing is conducted to certify that the aircraft meets all applicable safety and performance requirements of the government certifying agency. In the US, this is the Federal Aviation Administration (FAA); in Canada, Transport Canada (TC); in the United Kingdom (UK), the Civil Aviation Authority; and in the European Union, the European Aviation Safety Agency (EASA). Since commercial aircraft development is normally funded by the aircraft manufacturer and/or private investors, the certifying agency does not have a stake in the commercial success of the aircraft. These civil agencies are concerned with the aircraft's safety and that the pilot's flight manual accurately reports the aircraft's performance. The market will determine the aircraft's suitability to operators. Normally, the civil certification agency does not get involved in flight testing until the manufacturer has found and fixed any development issues and is ready to seek certification.

Military Aircraft

Military programs differ from commercial in that the government contracts with the aircraft manufacturer to design and build an aircraft to meet specific mission capabilities. These performance requirements are documented to the manufacturer in the aircraft specification and the details of the flight test program (among many other program requirements) are spelled out in the statement of work. In this case, the government is the customer and has a direct stake in the aircraft's ability to perform the mission. Since the government is funding the program, it is more involved in the aircraft design and testing from early-on. Often military test pilots and engineers are integrated as part of the manufacturer's flight test team, even before first flight. The final phase of the military aircraft flight test is the Operational Test (OT). OT is conducted by a government-only test team with the dictate to certify that the aircraft is suitable and effective to carry out the intended mission.

Flight testing of military aircraft is often conducted at military flight test facilities. The US Navy tests aircraft at Naval Air Station Patuxent River and the US Air Force at Edwards Air Force Base. The U.S. Air Force Test Pilot School and the U.S. Naval Test Pilot School are the programs designed to teach military test personnel. In the UK, most military flight testing is conducted by three organizations, the RAF, BAE Systems and QinetiQ. For minor upgrades the testing may be conducted by one of these three organizations in isolation, but major programs are normally conducted by a joint trials team (JTT), with all three organizations working together under the umbrella of an integrated project team (IPT) airspace.

Atmospheric Flight Testing of Launch Vehicles and Reusable Spacecraft

All launch vehicles, as well as a few reusable spacecraft, must necessarily be designed to deal with aerodynamic flight loads while moving through the atmosphere.

Many launch vehicles are flight tested, with rather more extensive data collection and analysis on the initial orbital launch of a particular launch vehicle design. Reusable spacecraft or reusable booster test programs are much more involved and typically follow the full envelope expansion paradigm of traditional aircraft testing. Previous and current test programs include the early drop tests of the Space Shuttle, the X-24B, Space Ship Two, Dream Chaser, and the SpaceX reusable launch system development program including the VTVL Grasshopper purpose-built booster-return test vehicle.

Flight Test Processes

Flight testing—typically as a class of non-revenue producing flight, although SpaceX has also done extensive flight tests on the post-mission phase of a returning booster flight on revenue launches—can be subject to the latter's statistically demonstrated higher risk of accidents or serious incidents. This is mainly due to the unknowns of a new aircraft or launch vehicle's handling characteristics and lack of established operating procedures, and can be exacerbated if test pilot training or experience of the flight crew is lacking For this reason, flight testing is carefully planned in three phases: preparation; execution; and analysis and reporting.

Preparation

For both commercial and military aircraft, as well as launch vehicles, flight test preparation begins well before the test vehicle is ready to fly. Initially what needs to be tested must be defined, from which the Flight Test Engineers prepare the test plan, which is essentially certain maneuvers to be flown (or systems to be exercised). Each single test is known as a Test Point. A full certification/qualification flight test program for a new aircraft will require testing for many aircraft systems and in-flight regimes; each is typically documented in a separate test plan. Altogether, a certification flight test program will consist of approximately 10,000 Test Points.

The document used to prepare a single test flight for an aircraft is known as a Test Card. This will consist of a description of the Test Points to be flown. The flight test engineer will try to fly similar Test Points from all test plans on the same flights, where practical. This allows the required data to be acquired in the minimum number of flight hours. The software used to control the flight test process is known as Flight Test Management Software, and supports the Flight Test Engineer in planning the test points to be flown as well as generating the required documentation.

Static pressure probe on the nose of a Sukhoi Superjet 100 prototype

Flight test pressure probes and water tanks in 747-8I prototype

Static pressure probe rig aboard Boeing 747-8I prototype; A long tube, rolled up inside the barrel, is connected to a probe which can be deployed far behind the tail of the aircraft

Once the flight test data requirements are established, the aircraft or launch vehicle is instrumented to record that data for analysis. Typical instrumentation parameters recorded during a flight test for a large aircraft are:

- Atmospheric (static) pressure and temperature;

- Dynamic ("total") pressure and temperature, measured at various positions around the fuselage;

- Structural loads in the wings and fuselage, including vibration levels;

- Aircraft attitude, angle of attack, and angle of sideslip;

- Accelerations in all six degrees of freedom, measured with accelerometers at different positions in the aircraft;

- Noise levels (interior and exterior);

- Internal temperature (in cabin and cargo compartments);

- Aircraft controls deflection (stick/yoke, rudder pedal, and throttle position);

- Engine performance parameters (pressure and temperature at various stages, thrust, fuel burn rate).

Specific calibration instruments, whose behavior has been determined from previous tests, may be brought on board to supplement the aircraft's in-built probes.

During the flight, these parameters are then used to compute relevant aircraft performance parameters, such as airspeed, altitude, weight, and center of gravity position.

During selected phases of flight test, especially during early development of a new aircraft, many parameters are transmitted to the ground during the flight and monitored by flight test and test support engineers, or stored for subsequent data analysis. This provides for safety monitoring and allows for both real-time and full-simulation analysis of the data being acquired.

Execution

When the aircraft or launch vehicle is completely assembled and instrumented, many hours of ground testing are conducted. This allows exploring multiple aspects: basic aircraft vehicle operation, flight controls, engine performance, dynamic systems stability evaluation, and provides a first look at structural loads. The vehicle can then proceed with its maiden flight, a major milestone in any aircraft or launch vehicle development program.

There are several aspects to a flight test program, among which:

- Handling qualities, which evaluates the aircraft's controllability and response to pilot inputs throughout the range of flight;

- Performance testing evaluates aircraft in relation to its projected abilities, such as speed, range, power available, drag, airflow characteristics, and so forth;

- Aero-elastic/flutter stability, evaluates the dynamic response of the aircraft controls and structure to aerodynamic (i.e. air-induced) loads;

- Avionics/systems testing verifies all electronic systems (navigation, communications, radars, sensors, etc.) perform as designed;

- Structural loads measure the stresses on the airframe, dynamic components, and controls to verify structural integrity in all flight regimes.

Testing that is specific to military aircraft includes :

- Weapons delivery, which looks at the pilot's ability to acquire the target using on-board systems and accurately deliver the ordnance on target;

- An evaluation of the separation of the ordnance as it leaves the aircraft to ensure there are no safety issues;

- air-to-air refueling;

- Radar/infrared signature measurement;

- Aircraft carrier operations.

Emergency situations are evaluated as a normal part of all flight test program. Examples are: engine failure during various phases of flight (takeoff, cruise, landing), systems failures, and controls degradation. The overall operations envelope (allowable gross weights, centers-of-gravity, altitude, max/min airspeeds, maneuvers, etc.) is established and verified during flight testing. Aircraft are always demonstrated to be safe beyond the limits allowed for normal operations in the Flight Manual.

Because the primary goal of a flight test program is to gather accurate engineering data, often on a design that is not fully proven, piloting a flight test aircraft requires a high degree of training and skill. As such, such programs are typically flown by a specially trained test pilot, the data is gathered by a flight test engineer, and often visually displayed to the test pilot and/or flight test engineer using flight test instrumentation.

Analysis and Reporting

It includes the analysis of a flight for certification.It analyze the internal and outer part of the flight by checking its all minute parts. Reporting includes the analyzed data result.

Introduction

Aircraft Performance has various missions such as Takeoff, Climb, Cruise, Acceleration, Deceleration, Descent, Landing and other Basic fighter maneuvers etc..

After the flight testing, the aircraft has to be certified according to their regulations like FAA's FAR, EASA's Certification Specifications (CS) and India's Air Staff Compliance and Requirements.

1. Flight Performance Evaluation and documentation

- Flight data processing includes filtering, bias correction and resolution along flight path (Trajectory).

- Analysis of mission segments from the flight test data.

- Estimation of thrust using Performance Cycle Deck (PCD).

- Calculation of In-flight thrust using In-Flight Thrust Deck (IFTD).

- Documentation of Flight performance with standard procedures.

- Validation and updating of Aircraft performance model.

2. Reduction of Flight performance to standard conditions

- Model estimation of Aircraft Performance with International Standard Atmosphere conditions (ISA).

- Non-standard (tested) conditions are studied by incorporating standard mass, altitude, speed, Throttle setting individually.

- The individual effects are added to the tested (non-standard) conditions to obtain the performance at International Standard Atmosphere conditions for Certification.

- For Takeoff and Landing, effect of Wind is considered too.

3. Preparation and Validation of Performance Charts for Operating Data Manual (ODM)

Performance charts allow a pilot to predict the takeoff, climb, cruise, and landing performance of an aircraft. These charts, provided by the manufacturer, are included in the AFM/POH. Information the manufacturer provides on these charts has been gathered from test flights conducted in a new aircraft, under normal operating conditions while using average piloting skills, and with the aircraft and engine in good working order. Engineers record the flight data and create performance charts based on the behavior of the aircraft during the test flights. By using these performance charts, a pilot can determine the runway length needed to take off and land, the amount of fuel to be used during flight, and the time required to arrive at the destination. It is important to remember that the data from the charts will not be accurate if the aircraft is not in good working order or when operating under adverse conditions. Always consider the necessity to compensate for the performance numbers if the aircraft is not in good working order or piloting skills are below average. Each aircraft performs differently and, therefore, has different performance numbers. Compute the performance of the aircraft prior to every flight, as every flight is different.

Every chart is based on certain conditions and contains notes on how to adapt the information for flight conditions. It is important to read every chart and understand how to use it. Read the instructions provided by the manufacturer. For an explanation on how to use the charts, refer to the example provided by the manufacturer for that specific chart.

The information manufacturers furnish is not standardized. Information may be contained in a table format, and other information may be contained in a graph format. Sometimes combined graphs incorporate two or more graphs into one chart to compensate for multiple conditions of flight. Combined graphs allow the pilot to predict aircraft performance for variations in density altitude, weight, and winds all on one chart. Because of the vast amount of information that can be extracted from this type of chart, it is important to be very accurate in reading the chart. A small error in the beginning can lead to a large error at the end.

The remainder of this section covers performance information for aircraft in general and discusses what information the charts contain and how to extract information from the charts by direct reading and interpolation methods. Every chart contains a wealth of information that should be used when flight planning. Examples of the table, graph, and combined graph formats for all aspects of flight will be discussed.

Interpolation Not all of the information on the charts is easily extracted. Some charts require interpolation to find the information for specific flight conditions. Interpolating information means that by taking the known information, a pilot can compute intermediate information. However, pilots sometimes round off values from charts to a more conservative figure. Using values that reflect slightly more adverse conditions provides a reasonable estimate of performance information and gives a slight margin of safety. The following illustration is an example of interpolating information from a takeoff distance chart.

- Model estimation for a wide range of atmospheric conditions, flight and engine parameters.

- Preparation and Validation of charts and tables from model estimation to predict the aircraft performance.

- This will enable the pilot to operate effectively and safely and do performance comparisons.

Flight Test Team

Flight test engineer's workstation aboard an Airbus A380 prototype

The make-up of the Flight Test Team will vary with the organization and complexity of the flight test program, however, there are some key players who are generally part of all flight test organizations. The leader of a flight test team is usually a flight test engineer (FTE) or possibly an experimental test pilot. Other FTEs or pilots could also be involved. Other team members would be the Flight Test Instrumentation Engineer, Instrumentation System Technicians, the aircraft maintenance department (mechanics, electrical techs, avionics technicians, etc.), Quality/Product Assurance Inspectors, the ground-based computing/data center personnel, plus logistics and administrative support. Engineers from various other disciplines would support the testing of their particular systems and analyze the data acquired for their specialty area.

Since many aircraft development programs are sponsored by government military services, military or government-employed civilian pilots and engineers are often integrated into the flight test team. The government representatives provide program oversight and review and approve data. Government test pilots may also participate in the actual test flights, possibly even on the first/maiden flight.

Cooper–Harper Rating Scale

The Cooper–Harper rating scale is a set of criteria used by test pilots and flight test engineers to evaluate the handling qualities of aircraft during flight test. The scale ranges from 1 to 10, with 1 indicating the best handling characteristics and 10 the worst. The criteria are evaluative and thus the scale is considered subjective.

Background

Development

After World War II, the various U.S. military branches sent different models of their operational aircraft to the Ames Aeronautical Laboratory located at the Moffett Federal Airfield in Mountain View, California for evaluation of the planes' flight performance and flying qualities. The laboratory was operated by NACA, the predecessor of NASA. Most of the flights were conducted by George Cooper, Bob Innis, and Fred Drinkwater and took place at the remote test site at the Crows Landing Naval Auxiliary Landing Field in the central valley area east of Moffett Field.

What may be the most important contribution of the flying qualities evaluation programs and experiments conducted on the variable stability aircraft at Ames was George Cooper's standardized system for rating an aircraft's flying qualities. Cooper developed his rating system over several years as a result of the need to quantify the pilot's judgment of an aircraft's handling in a fashion that could be used in the stability and control design process. This came about because of his perception of the value that such a system would have, and because of the encouragement of his colleagues in the United States and England who were familiar with his initial attempts.

Cooper's approach forced a specific definition of the pilot's task and of its performance standards. Furthermore, it accounted for the demands the aircraft placed on the pilot in accomplishing a given task to some specified degree of precision. The Cooper Pilot Opinion Rating Scale was initially published in 1957. After several years of experience gained in its application to many flight and flight simulator experiments, and through its use by the military services and aircraft industry, the scale was modified in collaboration with Robert (Bob) Harper of the Cornell Aeronautical Laboratory (Now Calspan) and became the Cooper-Harper Flying Qualities Rating Scale in 1969, a scale which remains the standard for measuring flying qualities.

Awards

In recognition of his many contributions to aviation safety, Cooper received the Admiral Luis de Florez Flight Safety Award in 1966 and the Richard Hansford Burroughs, Jr. Test Pilot Award in 1971. After he retired, both he and Bob Harper were selected by the American Institute of Aeronautics and Astronautics to reprise the Cooper-Harper Rating Scale in the 1984 Wright Brothers Lectureship in Aeronautics.

Other Scales

While the Cooper–Harper scale remains the only well-established scale for assessing aircraft flying qualities, its unidimensional format lacks diagnostic power and has also been criticised for exhibiting poor reliability. The Cranfield Aircraft Handling Qualities Rating Scale (CAHQRS), developed at Cranfield University's School of Engineering, is an evaluative system that is multidimensional. It was developed by combining concepts from two previously established scales, the NASA-TLX workload scale and the Cooper–Harper. A series of validation trials in an engineering flight simulator with a range of control laws showed that the CAHQRS was at least as effective as the Cooper–Harper scale. However, the CAHQRS also demonstrated greater diagnostic ability and reliability than the Cooper–Harper. This new scale's acceptance by the aerospace industry at large, though, remains to be demonstrated.

Scale

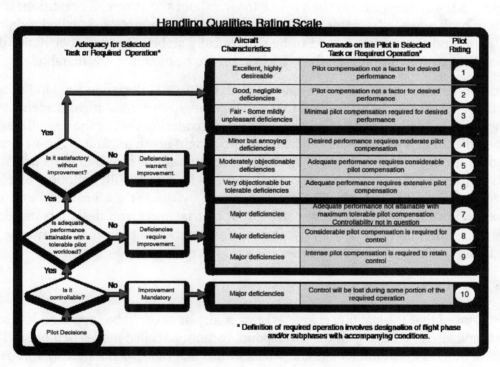

It is important to note that a Handling Qualities Rating (HQR) can not be assigned to an aircraft, as in "That aircraft is a HQR 5 aircraft." Any HQR that is assigned requires a well defined, repeatable task, a well trained pilot that is actively engaged in accomplishing that task, and an aircraft.

Flying Qualities

Handling qualities is one of the two principal regimes in the science of flight test (the other being performance). Handling qualities involves the study and evaluation of the stability and control characteristics of an aircraft. They have a critical bearing on the safety of flight and on the ease of controlling an airplane in steady flight and in maneuvers.

Relation to Stability

To understand the discipline of handling qualities, the concept of stability should be understood. Stability can be defined only when the vehicle is in trim; that is, there are no unbalanced forces or moments acting on the vehicle to cause it to deviate from steady flight. If this condition exists, and if the vehicle is disturbed, stability refers to the tendency of the vehicle to return to the trimmed condition. If the vehicle initially tends to return to a trimmed condition, it is said to be statically stable. If it continues to approach the trimmed condition without overshooting, the motion is called a subsidence. If the motion causes the vehicle to overshoot the trimmed condition, it may oscillate back and forth. If this oscillation damps out, the motion is called a damped oscillation and the vehicle is said to be dynamically stable. On the other hand, if the motion increases in amplitude, the vehicle is said to be dynamically unstable.

The theory of stability of airplanes was worked out by G. H. Bryan in England in 1904. This theory is essentially equivalent to the theory taught to aeronautical students today and was a remarkable intellectual achievement considering that at the time Bryan developed the theory, he had not even heard of the Wright brothers' first flight. Because of the complication of the theory and the tedious computations required in its use, it was rarely applied by airplane designers. Obviously, to fly successfully, pilotless airplanes had to be dynamically stable. The airplane flown by the Wright brothers, and most airplanes flown thereafter, were not stable, but by trial and error, designers developed a few planes that had satisfactory flying qualities. Many other airplanes, however, had poor flying qualities, which sometimes resulted in crashes.

Historical Development

Bryan showed that the stability characteristics of airplanes could be separated into longitudinal and lateral groups with the corresponding motions called modes of motion. These modes of motion were either aperiodic, which means that the airplane steadily approaches or diverges from a trimmed condition, or oscillatory, which means that the airplane oscillates about the trim condition. The longitudinal modes of a statically stable airplane following a disturbance were shown to consist of a long-period oscillation called the phugoid oscillation, usually with a period in seconds about one-quarter of the airspeed in miles per hour and a short-period oscillation with a period of only a few seconds. The lateral motion had three modes of motion: an aperiodic mode called the spiral mode that could be a divergence or subsidence, a heavily damped aperiodic mode called the roll subsidence, and a short-period oscillation, usually poorly damped, called the Dutch roll mode.

Some early airplane designers attempted to make airplanes that were dynamically stable, but it was found that the requirements for stability conflicted with those for satisfactory flying qualities. Meanwhile, no information was available to guide the designer as to just what characteristics should be incorporated to provide satisfactory flying qualities.

By the 1930s, there was a general feeling that airplanes should be dynamically stable, but some aeronautical engineers were starting to recognize the conflict between the requirements for stability and flying qualities. To resolve this question, Edward Warner, who was working as a consultant to the Douglas Aircraft Company on the design of the DC-4, a large four-engine transport airplane, made the first effort in the United States to write a set of requirements for satisfactory flying qualities. Dr. Warner, a member of the main committee of the NACA, also requested that a flight study be made to determine the flying qualities of an airplane along the lines of the suggested requirements. This study was conducted by Hartley A. Soulé of Langley. Entitled *Preliminary Investigation of the Flying Qualities of Airplanes*, Soulé's report showed several areas in which the suggested requirements needed revision and showed the need for more research on other types of airplanes. As a result, a program was started by Robert R. Gilruth with Melvin N. Gough as the chief test pilot.

Evaluation of Handling Qualities

The technique for the study of flying qualities requirements used by Gilruth was first to install instruments to record relevant quantities such as control positions and forces, airplane angular velocities, linear accelerations, airspeed, and altitude. Then a program of specified flight conditions and maneuvers was flown by an experienced test pilot. After the flight, data were transcribed from the records and the results were correlated with pilot opinion. This approach would be considered routine today, but it was a notable original contribution by Gilruth that took advantage of the flight recording instruments already available at Langley and the variety of airplanes available for tests under comparable conditions.

An important quantity in handling qualities measurements in turns or pull-ups is the variation of control force on the control stick or wheel with the value of acceleration normal to the flight direction expressed in g units. This quantity is usually called the force per g.

Relation to Spacecraft

Handling qualities are those characteristics of a flight vehicle that govern the ease and precision with which a pilot is able to perform a flying task. The way in which particular vehicle factors affect handling qualities has been studied in aircraft for decades, and reference standards for the handling qualities of both fixed-wing aircraft and rotary-wing aircraft have been developed and are now in common use. These standards define a subset of the dynamics and control design space that provides good handling qualities for a given vehicle type and flying task. A new generation of spacecraft now under development by NASA to replace the Space Shuttle and return astronauts to the Moon will have a manual control capability for several mission tasks, and the ease and precision with which pilots can execute these tasks will have an important effect on performance, mission risk and training costs. No reference standards currently exist for handling qualities of piloted spacecraft.

Drop Test

Orion test article being released during airborne drop test.

A drop test is a method of testing the in-flight characteristics of prototype or experimental aircraft and spacecraft by raising the test vehicle to a specific altitude and then releasing it. Test flights involving powered aircraft, particularly rocket-powered aircraft, may be referred to as drop launches due to the launch of the aircraft's rockets after release from its carrier aircraft.

In the case of unpowered aircraft, the test vehicle falls or glides after its release in an unpowered descent to a landing site. Drop tests may be used to verify the aerodynamic performance and flight dynamics of the test vehicle, to test its landing systems, or to evaluate survivability of a planned or crash landing. This allows the vehicle's designers to validate computer flight models, wind tunnel testing, or other theoretical design characteristics of an aircraft or spacecraft's design.

High-altitude drop tests may be conducted by carrying the test vehicle aboard a mothership to a target altitude for release. Low-altitude drop tests may be conducted by releasing the test vehicle from a crane or gantry.

Aircraft and Lifting-Body Testing

Carrier Landing Simulation Tests

The landing gear on aircraft used on aircraft carriers must be stronger than those on land-based aircraft, due to higher approach speeds and sink rates during carrier landings. As early as the 1940s, drop tests were conducted by lifting a carrier-based plane such as the Grumman F6F Hellcat to a height of ten feet and then dropped, simulating the impact of a landing at nineteen feet per second. The F6F was ultimately dropped from a height of twenty feet, demonstrating it could absorb twice the force of a carrier landing. Drop tests are still used in the development and testing of carrier-based aircraft; in 2010, the Lockheed Martin F-35C Lightning II underwent drop tests to simulate its maximum descent rate of 26.4 feet per second during carrier landings.

The X-38 research vehicle is released from *Balls 8*, NASA's B-52 mothership during a drop test. The pylon used to carry experimental vehicles is visible near the top of the photo, between the fuselage and inboard right engine.

Experimental Aircraft

Numerous experimental and prototype aircraft have been drop tested or drop launched. Many powered X-planes, including the Bell X-1, Bell X-2, North American X-15, Martin Marietta X-24A and X-24B, Orbital Sciences X-34, Boeing X-40, and NASA X-43A were specifically designed to be drop launched. Test articles of the unpowered NASA X-38 were also drop tested, from altitudes of up to 45,000 feet, in order to study its aerodynamic and handling qualities, autonomous flight capabilities, and deployment of its steerable parafoil.

Some experimental aircraft designed for airborne launches, such as the Northrop HL-10, have made both unpowered drop tests and powered drop launches. Prior to powered flights using its rocket engine, the HL-10 made 11 unpowered drop flights in order to study the handling qualities and stability of the lifting body in flight.

Balls 8 Mothership

Early experimental aircraft, such as the X-1 and X-2, were carried aboard modified B-29 and B-50 bombers. In the 1950s, the United States Air Force provided NASA with a B-52 bomber to be used as a mothership for the experimental X-15. Built in 1955, the B-52 was only the 10th to come off the assembly line, and was used by the Air Force for flight testing before turning it over to NASA. Flying with NASA tail number 008, the plane was nicknamed *Balls 8* by Air Force pilots, following a tradition of referring to aircraft numbered with multiple zeroes as "Balls" plus the final number.

Balls 8 received significant modifications in order to carry the X-15. A special pylon, designed to carry and release the X-15, was installed under the right wing between the fuselage and inboard engine. A notch was also cut out of one of the right wing's flaps so that the plane could accommodate the X-15's vertical tail. *Balls 8* was one of two such bombers modified to carry the X-15; while the other plane was retired in 1969 after the end of the X-15 program, NASA continued using *Balls 8* for drop tests until it was retired in 2004. During its 50-year career, *Balls 8* carried numerous experimental vehicles including the HL-10, X-24A, X-24B, X-38, and X-43A.

X-24B role in Space Shuttle Development

During the design of the Space Shuttle orbiter in the 1970s, engineers debated whether to design the orbiter to glide to an unpowered landing or equip the orbiter with pop-out jet engines in order to make a powered landing. While powered landing design required carrying the engines and jet fuel, adding weight and complexity to the orbiter, engineers began favoring the powered landing option. In response, NASA conducted unpowered drop tests of the X-24B to demonstrate the feasibility of landing a lifting-body aircraft in unpowered flight. In 1975, the X-24B aircraft was dropped from a *Balls 8* at an altitude of 45,000 feet above the Mojave Desert, and then ignited rocket engines to increase speed and propel it to 60,000 feet. Once the rocket engine cut off, the high-speed and high-altitude conditions permitted the X-24B to simulate the path of a Space Shuttle orbiter under post-atmospheric reentry conditions. The X-24B successfully made two unpowered precision landings at Edwards Air Force Base, demonstrating the feasibility of an unpowered lifting body design for the Space Shuttle. These successes convinced those in charge of the Space Shuttle program to commit to an unpowered landing design, which would save weight and increase the orbiter's payload capacity.

Enterprise being released by Shuttle Carrier Aircraft

Space Shuttle Enterprise

In 1977, a series of drop tests of the Space Shuttle *Enterprise* were conducted to test the Space Shuttle's flight characteristics. Because the Space Shuttle is designed to glide unpowered during its descent and landing, a series of drop tests using a test orbiter were used to demonstrate that the orbiter could be successfully controlled in unpowered flight. These drop tests, known as the Approach and Landing Test program, used a modified Boeing 747, known as the Shuttle Carrier Aircraft or SCA, to carry *Enterprise* to an altitude of 15,000 to 30,000 feet. After a series of captive-flight tests in which the orbiter was not released, five free-flight tests were performed in August through October 1977.

While free-flight tests of *Enterprise* involved the release of an unpowered aircraft from a powered aircraft, these tests were not typical of drop testing because the orbiter was actually carried and released from a position above the SCA. This arrangement was potentially dangerous because it placed *Enterprise* in free flight directly in front of the SCA's tail fin immediately after release. As a result, the "drop" was conducted by using a series of carefully planned maneuvers to minimize the risk of aircraft collision. Immediately after release, the *Enterprise* would climb to the right while the SCA performed a shallow dive to the left, allowing for quick vertical and horizontal separation between the two aircraft.

Dream Chaser

In summer 2013, Sierra Nevada Corporation plans to conduct drop tests of its Dream Chaser prototype commercial spaceplane. The unmanned first flight test will drop the Dream Chaser prototype from an altitude of 12,000 feet, where it is planned that the vehicle will autonomously fly to an unpowered at Dryden Flight Research Center.

Manned Capsule Testing

Drop tests of prototype manned space capsules may be done to test the survivability of landing, primarily by testing the capsule's descent characteristics and its post-reentry landing systems. These tests are typically carried out unmanned prior to any manned spaceflight testing.

Apollo Command Module

Orion test article after release from C-130 and separation from pallet

In 1963, North American Aviation built BP-19A, an unmanned boilerplate Apollo command module for use in drop testing. NASA conducted a series of tests in 1964 which involved dropping BP-19A from a C-133 Cargomaster in order to test the capsule's parachute systems prior to the start of manned testing of the Apollo spacecraft.

Orion Capsule

In 2011 and 2012, NASA conducted a series of short drop tests on the survivability of water landings in its Orion manned capsule by repeatedly dropping an Orion test vehicle into a large water basin. The tests simulated water landings at speeds varying from 7 mph to 50 mph by changing the height of the drop gantry above the basin. The range of landing velocities allowed NASA to simulate a range of possible entry and landing conditions during water landings.

In 2011 and 2012, NASA also conducted drop tests of the Orion test vehicle's parachute systems and land-based landing capabilities. In each test, the Orion spacecraft was dropped from a C-17 or C-130 cargo plane. For testing, the capsule is mounted on a pallet system and placed inside the cargo aircraft. Parachutes on the pallet are used to pull the pallet and capsule out of the rear of the aircraft; the capsule then separates from the pallet and begins its free fall descent.

On March 4, 2012, a C-17 dropped an Orion test article from an altitude of 25,000 feet. The capsule's parachutes successfully deployed between 15,000 and 20,000 feet, slowing the spacecraft to a landing on ground in the Arizona desert. The capsule landed at a speed of 17 mph, well below the designed maximum touchdown speed.

Boeing CST-100

In September 2011, Boeing conducted a series of drop tests, carried out in the Mojave Desert of southeast California, to validate the design of the CST-100 capsule's parachute and airbag cushioning landing systems. The airbags are located underneath the heat shield of the CST-100, which is designed to be separated from the capsule while under parachute descent at about 5,000 feet (1,500 m) altitude. The tests were carried out at ground speeds between 10 and 30 miles per hour (16 and 48 km/h) in order to simulate cross wind conditions at the time of landing. Bigelow Aerospace built the mobile test rig and conducted the tests.

In April 2012, Boeing conducted another drop test of its CST-100 prototype space capsule in order to test the capsule's landing systems. The test vehicle was raised by helicopter to an altitude of 11,000 feet and then released; the capsule's three main parachutes then deployed successfully and slowed the capsule's descent. Immediately prior to landing, the capsule's six airbags inflated underneath the capsule in order to absorb some of the impact energy from landing. Similar drop tests are planned in order to conduct additional airbag testing, as well as drogue chute and heat shield jettison tests.

Helicopter Testing

In 2009 and 2010, NASA conducted a pair of drop tests to study the survivability of helicopter crashes. Using an MD 500 helicopter donated by the U.S. Army, NASA dropped the helicopter at an angle from an altitude of 35 feet to simulate a hard helicopter landing. Sophisticated crash test dummies with simulated internal organs were located inside the helicopter and used to assess internal injuries from such a crash. Due to extensive damage to the test helicopter after the second test, no third test was planned.

Electrically Powered Spacecraft Propulsion

An electrically powered spacecraft propulsion system uses electrical energy to change the velocity of a spacecraft. Most of these kinds of spacecraft propulsion systems work by electrically expelling propellant (reaction mass) at high speed, but electrodynamic tethers work by interacting with a planet's magnetic field.

Electric thrusters typically use much less propellant than chemical rockets because they have a higher exhaust speed (operate at a higher specific impulse) than chemical rockets. Due to limited electric power the thrust is much weaker compared to chemical rockets, but electric propulsion can provide a small thrust for a long time. Electric propulsion can achieve high speeds over long periods and thus can work better than chemical rockets for some deep space missions.

Electric propulsion is now a mature and widely used technology on spacecraft. Russian satellites have used electric propulsion for decades. As of 2013, over 200 spacecraft operated throughout the solar system use electric propulsion for station keeping, orbit raising, or primary propulsion. In the future, the most advanced electric thrusters may be able to impart a Delta-v of 100 km/s, which is enough to take a spacecraft to the outer planets of the Solar System (with nuclear power), but is insufficient for interstellar travel. Also, an electro-rocket with an external power source (transmissible through laser on the solar panels) has a theoretical possibility for interstellar flight. However, electric propulsion is not a method suitable for launches from the Earth's surface, as the thrust for such systems is too weak.

History

The idea of electric propulsion for spacecraft dates back to 1911, introduced in a publication by Konstantin Tsiolkovsky. Earlier, Robert Goddard had noted such a possibility in his personal notebook.

The first in the world designed and tested electric propulsion was in 1929-1931 in Leningrad. Already in 1950 at the initiative of S.P. Korolev, I.V. Kurchatov and L.A. Artsimovich it adopted a program of research and development of various electrical rocket engines.

Electrically powered propulsion with a nuclear reactor was considered by Dr. Tony Martin for interstellar Project Daedalus in 1973, but the novel approach was rejected because of very low thrust, the heavy equipment needed to convert nuclear energy into electricity, and as a result a small acceleration, which would take a century to achieve the desired speed.

The demonstration of electric propulsion was an ion engine carried on board the SERT-1 (Space Electric Rocket Test) spacecraft, launched on 20 July 1964 and it operated for 31 minutes. A follow-up mission launched on 3 February 1970, SERT-2, carried two ion thrusters, one operated for more than five months and the other for almost three months.

By the early 2010s, many satellite manufacturers were offering electric propulsion options on their satellites—mostly for on-orbit attitude control—while some commercial communication satellite operators were beginning to use them for geosynchronous orbit insertion in place of traditional chemical rocket engines.

Types

Ion and Plasma Drives

This type of rocket-like reaction engine uses electric energy to obtain thrust from propellant carried with the vehicle. Unlike rocket engines, these kinds of engines do not necessarily have rocket nozzles, and thus many types are not considered true rockets.

Electric propulsion thrusters for spacecraft may be grouped in three families based on the type of force used to accelerate the ions of the plasma:

Electrostatic

If the acceleration is caused mainly by the Coulomb force (i.e. application of a static electric field in the direction of the acceleration) the device is considered electrostatic.

- Gridded ion thruster
    - NASA Solar Technology Application Readiness (NSTAR)
    - HiPEP
    - Radiofrequency ion thruster
- Hall effect thruster
    - SPT – Stationary Plasma Thruster

  o   TAL – Thruster with Anode Layer

- Colloid ion thruster

- Field Emission Electric Propulsion

- Nano-particle field extraction thruster

- Contact ion thruster

- Plasma separator ion thruster

- Radioisotopic ion thruster

Electrothermal

The electrothermal category groups the devices where electromagnetic fields are used to generate a plasma to increase the temperature of the bulk propellant. The thermal energy imparted to the propellant gas is then converted into kinetic energy by a nozzle of either solid material or magnetic fields. Low molecular weight gases (e.g. hydrogen, helium, ammonia) are preferred propellants for this kind of system.

An electrothermal engine uses a nozzle to convert the heat of a gas into the linear motion of its molecules so it is a true rocket even though the energy producing the heat comes from an external source.

Performance of electrothermal systems in terms of specific impulse (Isp) is somewhat modest (500 to ~1000 seconds), but exceeds that of cold gas thrusters, monopropellant rockets, and even most bipropellant rockets. In the USSR, electrothermal engines were used since 1971; the Soviet "Meteor-3", "Meteor-Priroda", "Resurs-O" satellite series and the Russian "Elektro" satellite are equipped with them. Electrothermal systems by Aerojet (MR-510) are currently used on Lockheed Martin A2100 satellites using hydrazine as a propellant.

- Arcjet

- Microwave arcjet

- Resistojet

Electromagnetic

If ions are accelerated either by the Lorentz force or by the effect of an electromagnetic fields where the electric field is not in the direction of the acceleration, the device is considered electromagnetic.

- Electrodeless plasma thruster

- MPD thruster

- Pulsed inductive thruster

- Pulsed plasma thruster

- Helicon Double Layer Thruster

- Variable specific impulse magnetoplasma rocket (VASIMR)

Non-ion Drives

Photonic

Photonic drive does not expel matter for reaction thrust, only photons.

Electrodynamic Tether

Electrodynamic tethers are long conducting wires, such as one deployed from a tether satellite, which can operate on electromagnetic principles as generators, by converting their kinetic energy to electric energy, or as motors, converting electric energy to kinetic energy. Electric potential is generated across a conductive tether by its motion through the Earth's magnetic field. The choice of the metal conductor to be used in an electrodynamic tether is determined by a variety of factors. Primary factors usually include high electrical conductivity, and low density. Secondary factors, depending on the application, include cost, strength, and melting point.

Unconventional

The principle of action of these theoretical devices is not well explained by the currently-understood laws of physics.

- Quantum Vacuum Plasma Thruster

- EM Drive or Cannae Drive

Steady vs. Unsteady

Electric propulsion systems can also be characterized as either steady (continuous firing for a prescribed duration) or unsteady (pulsed firings accumulating to a desired impulse). However, these classifications are not unique to electric propulsion systems and can be applied to all types of propulsion engines.

Dynamic Properties

Electrically powered rocket engines provide lower thrust compared to chemical rockets by several orders of magnitude because of the limited electrical power possible to provide in a spacecraft. A chemical rocket imparts energy to the combustion products directly, whereas an electrical system requires several steps. However, the high velocity and lower reaction mass expended for the same thrust allows electric rockets to run for a long time. This differs from the typical chemical-powered spacecraft, where the engines run only in short intervals of time, while the spacecraft mostly follows an inertial trajectory. When near a planet, low-thrust propulsion may not offset the gravitational attraction of the planet. An electric rocket engine cannot provide enough thrust to lift the vehicle from a planet's surface, but a low thrust applied for a long interval can allow a spacecraft to maneuver near a planet.

Airframe

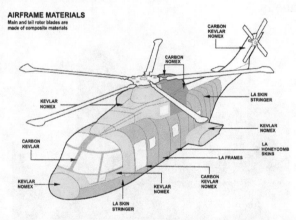

Airframe diagram for an AgustaWestland AW101 helicopter

The airframe of an aircraft is its mechanical structure. It is typically considered to include fuselage, wings and undercarriage and exclude the propulsion system. Airframe design is a field of aerospace engineering that combines aerodynamics, materials technology and manufacturing methods to achieve balances of performance, reliability and cost.

History

Modern airframe history began in the United States when a 1903 wood biplane made by Orville and Wilbur Wright showed the potential of fixed-wing designs. Many early developments were spurred by military needs during World War I. Well known aircraft from that era include the Dutch designer Anthony Fokker's combat aircraft for the German Empire's *Luftstreitkräfte*, and U.S. Curtiss flying boats and the German/Austrian Taube monoplanes. These used hybrid wood and metal structures. During the war, German engineer Hugo Junkers pioneered practical all-metal airframes as early as late 1915 with the Junkers J 1 and developed further with lighter weight duralumin in the airframe of the Junkers D.I of 1918, whose techniques were adopted almost unchanged after the war by both American engineer William Bushnell Stout and Soviet aerospace engineer Andrei Tupolev. Commercial airframe development during the 1920s and 1930s focused on monoplane designs using radial piston engines. Many, such as the Ryan model flown across the Atlantic by Charles Lindbergh in 1927, were produced as single copies or in small quantity. William Stout's designs for the all-metal Ford 4-AT and 5-AT trimotors, Andrei Tupolev's designs in Joseph Stalin's Soviet Union for a series of all-metal aircraft of steadily increasing size, culminating in the enormous, eight-engined *Maksim Gorky* (the largest aircraft of its era), and with Donald Douglas' firm's development of the iconic Douglas DC-3 twin-engined airliner, were among the most successful designs to emerge from the era through the use of all-metal airframes. The original Junkers corrugated duralumin-covered airframe philosophy culminated in the 1932-origin Junkers Ju 52 trimotor airliner, used throughout World War II by the Nazi German Luftwaffe for transport and paratroop needs.

Wellington Mark X showing the geodesic airframe construction and the level of punishment it could withstand while maintaining airworthiness

During World War II, military needs again dominated airframe designs. Among the best known were the US Douglas C-47, Boeing B-17, North American B-25 and Lockheed P-38, and British Vickers Wellington that used a geodesic construction method, and Avro Lancaster, all revamps of original designs from the 1930s. The wooden composite construction high performance fighter-bomber de Havilland Mosquito was developed during the war. The first jets were produced during the war but not made in large quantity. The Boeing B-29 was designed to be a high altitude bomber, the first with a pressurised fuselage.

Postwar commercial airframe design focused on larger capacities, on turboprop engines, and then on jet (turbojet, later turbofan) engines. The generally higher speeds and stresses of turboprops and jets were major challenges. Newly developed aluminum alloys with copper, magnesium and zinc were critical to these designs. The Lockheed L-188 turboprop, first flown in 1957, used some of these materials and became a costly lesson in controlling vibration and planning around metal fatigue.

DH106 Comet 3 G-ANLO demonstrating at the 1954 Farnborough Airshow

The de Havilland Comet was the world's first commercial jet airliner to reach production. It first flew in 1949 and was considered a landmark in British aeronautical design. After introduction into commercial service, early Comet models suffered from catastrophic airframe metal fatigue, causing a string of well-publicised accidents. The Royal Aircraft Establishment investigation at Farnborough, founded the science of aircraft crash reconstruction. Over 3000 cycles of pressurisation later, in a specially constructed pressure chamber, airframe failure was found to be due to stress concentration, a consequence of the square shaped windows. The windows had been engineered to be glued and riveted, but had been punch riveted only. Unlike drill riveting, the imperfect nature of the hole created by punch riveting may cause the start of fatigue cracks around the rivet.

Eventually Boeing in the U.S. and Airbus in Europe became the dominant assemblers of large airframes, known as wide-body aircraft. Numerous manufacturers in Europe, North America and South America took over markets for airframes designed to carry 100 or fewer passengers. Many manufacturers produce airframe components.

Present and Future

Rough interior of a Boeing 747 airframe

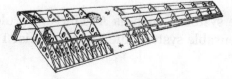

Wing structure with ribs and one spar

Four major eras in commercial airframe production stand out: all-aluminum structures beginning in the 1920s and directly inspired by Hugo Junkers all-metal designs from as far back as 1915, high-strength alloys and high-speed airfoils beginning in the 1940s, long-range designs and improved efficiencies beginning in the 1960s, and composite material construction beginning in the 1980s, partly pioneered by Burt Rutan's designs. In the latest era, Boeing has claimed a lead, designing its new 787 jetliner with a one-piece carbon-fiber fuselage, said to replace "1,200 sheets of aluminum and 40,000 rivets." The Airbus A380 is also built with a large proportion of composite material. Cirrus Aircraft's SR20 design, certified in 1998, was the first general aviation aircraft manufactured with all-composite construction, followed by several other light aircraft in the 2000s.

Airframe production has become an exacting process. Manufacturers operate under strict quality control and government regulations. Departures from established standards become objects of major concern. The crash on takeoff of an Airbus A300 in 2001, after its tail assembly broke away from the fuselage, called attention to operation, maintenance and design issues involving composite materials that are used in many recent airframes. The A300 had experienced other structural problems but none of this magnitude. The incident bears comparison with the 1959 Lockheed L-188 crash in showing difficulties that the airframe industry and its airline customers can experience when adopting new technology.

Reusable Launch System

A reusable launch system (RLS, or reusable launch vehicle, RLV) is a launch system which is capable of launching a payload into space more than once. This contrasts with expendable launch systems, where each launch vehicle is launched once and then discarded.

No completely reusable orbital launch system has ever been created; however, several partially reusable launch systems have existed. The Space Shuttle was partially reusable: the orbiter, which included the Space Shuttle main engines, and the two solid rocket boosters, were reused after several months of refitting work for each launch. However, the external tank and launch vehicle load frame were discarded after each flight.

The Falcon 9 rocket is designed to have a reusable first stage; several of these stages have been safely returned to land after launch. However, as of 2016, none of these first stages have yet been reused.

Several partially reusable systems, such as Adeline and Vulcan, are currently under development; one fully reusable system, the Mars Colonial Transporter, is also under development.

Orbital RLVs are thought to provide the possibility of low cost and highly reliable access to space. However, reusability implies weight penalties such as non-ablative reentry shielding and possibly a stronger structure to survive multiple uses, and given the lack of experience with these vehicles, the actual costs and reliability are yet to be seen.

History

ROMBUS

Aerospaceplane 1

In the first half of the twentieth century, popular science fiction often depicted space vehicles as either single-stage reusable rocket ships which could launch and land vertically (SSTO VTVL), or single-stage reusable rocket planes which could launch and land horizontally (SSTO HTHL).

The realities of early engine technology with low specific impulse or insufficient thrust-to-weight ratio to escape Earth's gravity well, compounded by construction materials without adequate performance (strength, stiffness, heat resistance) and low weight, seemingly rendered that original single-stage reusable vehicle vision impossible.

However, advances in materials and engine technology have rendered this concept potentially feasible.

Before VTVL SSTO designs came the partially reusable multi-stage NEXUS launcher by Krafft Arnold Ehricke. The pioneer in the field of VTVL SSTO, Philip Bono, worked at Douglas. Bono proposed several launch vehicles including: ROOST, ROMBUS, Ithacus, Pegasus and SASSTO. Most of his vehicles combined similar innovations to achieve SSTO capability. Bono proposed:

- Plug nozzle engines to retain high specific impulse at all altitudes.

- Base first reentry which allowed the reuse of the engine as a heat shield, lowering required heat shield mass.

- Use of spherical tanks and stubby shape to reduce vehicle structural mass further.

- Use of drop tanks to increase range.

- Use of in-orbit refueling to increase range.

Bono also proposed the use of his vehicles for space launch, rapid intercontinental military transport (Ithacus), rapid intercontinental civilian transport (Pegasus), even Moon and Mars missions (Project Selena, Project Deimos).

In Europe, Dietrich Koelle, inspired by Bono's SASSTO design, proposed his own VTVL vehicle named BETA.

Before HTHL SSTO designs came Eugen Sänger and his Silbervogel ("Silverbird") suborbital skip bomber. HTHL vehicles which can reach orbital velocity are harder to design than VTVL due to their higher vehicle structural weight. This led to several multi-stage prototypes such as a suborbital X-15. Aerospaceplane being one of the first HTHL SSTO concepts. Proposals have been made to make such a vehicle more viable including:

- Rail boost (e.g. 270 m/s at 3000 m on a mountain allowing 35% less SSTO take-off mass for a given payload in one NASA study)

- Use of lifting body designs to reduce vehicle structural mass.

- Use of in-flight refueling.

Other launch system configuration designs are possible such as horizontal launch with vertical landing (HTVL) and vertical launch with horizontal landing (VTHL). One of the few HTVL vehicles is the 1960s concept spacecraft Hyperion SSTO, designed by Philip Bono. X-20 Dyna-Soar is an early example of a VTHL design, while the HL-20 and X-34 are examples from the 1990s. As of February 2010, the VTHL X-37 has completed initial development and flown an initial classified orbital mission of over seven months duration. Currently proposed VTHL manned spaceplanes include the Dream Chaser and Prometheus, both circa 2010 concept spaceplanes proposed to NASA under the CCDev program.

The late 1960s saw the start of the Space Shuttle design process. From an initial multitude of ideas a two-stage reusable VTHL design was pushed forward that eventually resulted in a reusable orbiter payload spacecraft and reusable solid rocket boosters. The external tank and the launch vehicle load frame were discarded, and the parts that were reusable took a 10,000-person group nine months to refurbish for flight. So the space shuttle ended up costing a billion dollars per flight. Early studies from 1980 and 1982 proposed in-space uses for the tank to be re-used in space for various applications but NASA never pursued those options beyond the proposal stage.

During the 1970s further VTVL and HTHL SSTO designs were proposed for solar power satellite and military applications. There was a VTVL SSTO study by Boeing. HTHL SSTO designs included the Rockwell Star-Raker and the Boeing HTHL SSTO study. However the focus of all space launch funding in the United States on the Shuttle killed off these prospects. The Soviet Union followed suit with Buran. Others preferred expendables for their lower design risk, and lower design cost.

Eventually the Shuttle was found to be expensive to maintain, even more expensive than an expendable launch system would have been. The cancellation of a Shuttle-Centaur rocket after the loss of Challenger also caused an hiatus that would make it necessary for the United States military to scramble back towards expendables to launch their payloads. Many commercial satellite customers had switched to expendables even before that, due to unresponsiveness to customer concerns by the Shuttle launch system.

In 1986 President Ronald Reagan called for an airbreathing scramjet plane to be built by the year 2000, called NASP/X-30 that would be capable of SSTO. Based on the research project copper canyon the project failed due to severe technical issues and was cancelled in 1993.

This research may have inspired the British HOTOL program, which rather than airbreathing to high hypersonic speeds as with NASP, proposed to use a precooler up to Mach 5.5. The program's funding was canceled by the British government when the research identified some technical risks as well as indicating that that particular vehicle architecture would only be able to deliver a relatively small payload size to orbit.

When the Soviet Union collapsed in the early nineties, the cost of Buran became untenable. Russia has only used pure expendables for space launch since.

The 1990s saw interest in developing new reusable vehicles. The military Strategic Defense Initiative ("Star Wars") program "Brilliant Pebbles" required low cost, rapid turnaround space launch. From this requirement came the McDonnell Douglas Delta Clipper VTVL SSTO proposal. The DC-X prototype for Delta Clipper demonstrated rapid turnaround time and that automatic computer control of such a vehicle was possible. It also demonstrated it was possible to make a reusable space launch vehicle which did not require a large standing army to maintain like the Shuttle.

In mid-1990, further British research and major reengineering to avoid deficiencies of the HOTOL design led to the far more promising Skylon design, with much greater payload.

From the commercial side, large satellite constellations such as Iridium satellite constellation were proposed which also had low cost space access demands. This fueled a private launch industry, including partially reusable vehicle players, such as Rocketplane Kistler, and reusable vehicle players such as Rotary Rocket.

The end of that decade saw the implosion of the satellite constellation market with the bankruptcy of Iridium. In turn the nascent private launch industry collapsed. The fall of the Soviet Union eventually had political ripples which led to a scaling down of ballistic missile defense, including the demise of the "Brilliant Pebbles" program. The military decided to replace their aging expendable launcher workhorses, evolved from ballistic missile technology, with the EELV program. NASA proposed riskier reusable concepts to replace Shuttle, to be demonstrated under the X-33 and X-34 programs.

The 21st century saw rising costs and teething problems lead to the cancellation of both X-33 and X-34. Then the Space Shuttle Columbia disaster and another grounding of the fleet. The Shuttle design was now over 20 years old and in need of replacement. Meanwhile, the military EELV program churned out a new generation of better expendables. The commercial satellite market is depressed due to a glut of cheap expendable rockets and there is a dearth of satellite payloads.

Against this backdrop came the Ansari X Prize contest, inspired by the aviation contests made in the early 20th century. Many private companies competed for the Ansari X Prize, the winner being Scaled Composites with their reusable HTHL SpaceShipOne. It won the ten million dollars, by reaching 100 kilometers in altitude twice in a two-week period with the equivalent of three people on board, with no more than ten percent of the non-fuel weight of the spacecraft replaced between flights. While SpaceShipOne is suborbital like the X-15, some hope the private sector can eventually develop reusable orbital vehicles given enough incentive. SpaceX is a recent player in the private launch market succeeding in converting its Falcon 9 expendable launch vehicle into a partially reusable vehicle by returning the first stage for reuse.

On 23 November 2015, Blue Origin New Shepard rocket became the first proven Vertical Take-off Vertical Landing (VTVL) rocket which can reach space, by passing Kármán line (100 kilometres), reaching 329,839 feet (100.5 kilometers). Previous VTVL record was in 1994, the McDonnell Douglas DC-X ascended to an altitude of about 3.1 kilometers before successfully landing.

Reusability Concepts

Single Stage

There are two approaches to Single stage to orbit or SSTO. The rocket equation says that an SSTO vehicle needs a high mass ratio. Mass ratio is defined as the mass of the fully fueled vehicle divided by the mass of the vehicle when empty (zero fuel weight, ZFW).

One way to increase the mass ratio is to reduce the mass of the empty vehicle by using very lightweight structures and high efficiency engines. This tends to push up maintenance costs as component reliability can be impaired, and makes reuse more expensive to achieve. The margins are so small with this approach that there is uncertainty whether such a vehicle would be able to carry any payload into orbit.

Two or more Stages to Orbit

Two stage to orbit requires designing and building two independent vehicles and dealing with the interactions between them at launch. Usually the second stage in launch vehicle is 5-10 times smaller than the first stage, although in **biamese** and **triamese** approaches each vehicle is the same size.

In addition, the first stage needs to be returned to the launch site for it to be reused. This is usually proposed to be done by flying a compromise trajectory that keeps the first stage above or close to the launch site at all times, or by using small airbreathing engines to fly the vehicle back, or by recovering the first stage downrange and returning it some other way (often landing in the sea, and returning it by ship.) Most techniques involve some performance penalty; these can require the first stage to be several times larger for the same payload, although for recovery from downrange these penalties may be small.

The second stage is normally returned after flying one or more orbits and reentering.

Biamese & Triamese (Crossfeed)

Two or three similar stages are stacked side by side, and burn in parallel. Using crossfeed, the fuel tanks of the orbital stage are kept full, while the tank(s) in the booster stage(s) are used to run engines in the booster stage(s) and orbital stage. Once the boosters run dry, they are ejected, and (typically) glide back to a landing. The advantage to this is that the mass ratios of the individual stages is vastly reduced due to the way cross feed modifies the rocket equation. $Isp*g*\ln(2MR^2/MR+1)$ & $Isp*g*\ln(3MR^2/MR+2)$ respectively. With hydrogen engines, a triamese only needs an MR of 5, as opposed to an MR of 10 for a single-stage equivalent vehicle.

A criticism of this approach is that designing separate orbiter and boosters, or a single vehicle that could do both, would compromise performance, safety, and possible cost savings. Compromising maximum performance to reduce cargo cost however, is the point of the triamese approach. Stacking two or three winged vehicles can also be challenging. Optimistically, the lower mass ratios would translate to lower overall R&D costs, even if two different stage designs. While many aerospace designs have successfully been modified far beyond the original designers intentions (Boeing's 747 is perhaps the best example) the slow and painful birth of the F-35 family demonstrates that it is not always a guarantee of such flexibility.

Horizontal Landing

Scaled Composites SpaceShipOne used horizontal landing after being launched from a carrier airplane

In this case the vehicle requires wings and undercarriage (unless landing at sea). This typically requires about 9-12% of the landing vehicle to be wings; which in turn implies that the takeoff weight is higher and/or the payload smaller.

Concepts such as lifting bodies attempt to deal with the somewhat conflicting issues of reentry, hypersonic and subsonic flight; as does the delta wing shape of the Space Shuttle.

Vertical Landing

McDonnell Douglas DC-X used vertical takeoff and vertical landing

Parachutes could be used to land vertically, either at sea, or with the use of small landing rockets, on land (as with Soyuz). McDonnell Douglas DC-X ascended to an altitude of about 3.1 kilometers before successfully landing.

Alternatively rockets could be used to softland the vehicle on the ground from the subsonic speeds reached at low altitude. This typically requires about 10% of the landing weight of the vehicle to be propellant.

A slightly different approach to vertical landing is to use an autogyro or helicopter rotor. This requires perhaps 2-3% of the landing weight for the rotor.

SpaceX's grasshopper rocket, a 10-story Vertical Takeoff Vertical Landing (VTVL) vehicle, became the first reusable rocket designed to test the technologies needed to return a rocket back to Earth intact. While most rockets are designed to burn up in the atmosphere during reentry, SpaceX's rockets are being designed to return to the launch pad for a vertical landing.

Blue Origin New Shepard rocket became the first proven rocket which can do vertical landing after reaching space, by passing Kármán line (100 kilometres).

SpaceX's Falcon 9 rocket became the first orbital rocket to vertically land its first stage on the ground, after propelling its second stage and payload to a suborbital trajectory, where it would continue on to orbit.

Horizontal Takeoff

XCOR Aerospace EZ-Rocket used horizontal takeoff and landing using a standard airport runway

The vehicle needs wings to take off. For reaching orbit, a 'wet wing' would often need to be used where the wing contains propellant. Around 9-12% of the vehicle takeoff weight is perhaps tied up in the wings.

Vertical Takeoff

This is the traditional takeoff regime for pure rocket vehicles. Rockets are good for this regime, since they have a very high thrust/weight ratio (~100).

Airbreathing

Airbreathing approaches use the air during ascent for propulsion. The most commonly proposed approach is the scramjet, but turborocket, Liquid Air Cycle Engine (LACE) and precooled jet engines have also been proposed.

In all cases the highest speed that an airbreathing engine can reach is far short of orbital speed (about Mach 15 for Scramjets and Mach 5-6 for the other engine designs), and rockets would be used for the remaining 10-20 Mach into orbit.

The thermal situation for airbreathers (particularly scramjets) can be awkward; normal rockets fly steep initial trajectories to avoid drag, whereas scramjets would deliberately fly through relatively thick atmosphere at high speed generating enormous heating of the airframe. The thermal situation for the other airbreathing approaches is much more benign, although is not without its challenges.

Propellant

Hydrogen Fuel

Hydrogen is often proposed since it has the highest exhaust velocity. However tankage and pump weights are high due to insulation and low propellant density; and this wipes out much of the advantage.

Still, the 'wet mass' of a hydrogen fuelled stage is lighter than an equivalent dense stage with the same payload and this can permit usage of wings, and is good for second stages.

Dense Fuel

Dense fuel is sometimes proposed since, although it implies a heavier vehicle, the specific tankage and pump mass is much improved over hydrogen. Dense fuel is usually suggested for vertical takeoff vehicles, and is compatible with horizontal landing vehicles, since the vehicle is lighter than an equivalent hydrogen vehicle when empty of propellant. Non-cryogenic dense fuels also permit the storage of fuel in wing structures. Projects have been underway to densify existing fuel types through various techniques. These include slush technologies for cryogenics like hydrogen and propane. Another densifying method has been studied that would also increase the specific impulse of fuels. Adding finely powdered carbon, aluminum, titanium, and boron to hydrogen and kerosene have been studied. These additives increase the specific impulse (Isp) but also the density of the fuel. For instance, the French ONERA missile program tested boron with kerosene in gelled slurries, as well as embedded in paraffin, and demonstrated increases in volumetric specific impulse of between 20-100%.

Tripropellant

Dense fuel is optimal early on in a flight, since the thrust to weight of the engines is better due to higher density; this means the vehicle accelerates more quickly and reaches orbit sooner, reducing gravity losses.

However, for reaching orbital speed, hydrogen is a better fuel, since the high exhaust velocity and hence lower propellant mass reduces the take off weight.

Therefore, tripropellant vehicles start off burning with dense fuel and transition to hydrogen. (In a sense the Space Shuttle does this with its combination of solid rockets and main engines, but tripropellant vehicles usually carry their engines to orbit.)

Propellant Costs

As with all current launch vehicles, propellant costs for a rocket are much lower than the costs of the hardware. However, for reusable vehicles if the vehicles are successful, then the hardware is reused many times and this would bring the costs of the hardware down. In addition, reusable vehicles are frequently heavier and hence less propellant efficient, so the propellant costs could start to multiply up to the point where they become significant.

Launch Assistance/non Rocket Space Launch

Since rocket delta-v has a non linear relationship to mass fraction due to the rocket equation, any small reduction in delta-v gives a relatively large reduction in the required mass fraction; and starting a mission at higher altitude also helps.

Many systems have proposed the use of aircraft to gain some initial velocity and altitude; either by towing, carrying or even simply refueling a vehicle at altitude.

Various other launch assists have been proposed, such as ground-based sleds, or maglev systems, high altitude (80 km) maglev systems such as launch loops, to more exotic systems such as tether propulsion systems to catch the vehicle at high altitude, or even Space Elevators.

Reentry Heat Shields

Robert Zubrin has said that as a rough rule of thumb, 15% of the landed weight of a vehicle needs to be aerobraking reentry shielding.

Reentry heat shields on these vehicles are often proposed to be some sort of ceramic and/or carbon-carbon heat shields, or occasionally metallic heat shields (possibly using water cooling or some sort of relatively exotic rare earth metal.) Some shields would be single-use ablatives, discarded after reentry.

A newer Thermal Protection System (TPS) technology was first developed for use in steering fins on ICBM MIRVs. Given the need for such warheads to reenter the atmosphere swiftly and retain hypersonic velocities to sea level, researchers developed what are known as SHARP materials, typically hafnium diboride and zirconium diboride, whose thermal tolerance exceeds 3600 C. SHARP equipped vehicles can fly at Mach 11 at 30 km altitude and Mach 7 at sea level. The sharp-edged geometries permitted with these materials also eliminates plasma shock wave interference in radio communications during reentry. SHARP materials are very robust and would not require constant maintenance, as is the case with technologies like silica tiles, used on the Space Shuttle, which account for over half of that vehicles maintenance costs and turnaround time. The maintenance savings alone are thus a major factor in favor of using these materials for a reusable launch vehicle, whose raison d'etre is high flight rates for economical launch costs.

Weight Penalty

The weight of a reusable vehicle is almost invariably higher than an expendable that was made with the same materials, for a given payload.

R&D

The research & development costs of reusable vehicle are expected to be higher, because making a vehicle reusable implies making it robust enough to survive more than one use, which adds to the testing required. Increasing robustness is most easily done by adding weight; but this reduces performance and puts further pressure on the R&D to recoup this in some other way.

These extra costs must be recouped; and this pushes up the average cost of the vehicle.

Maintenance

Reusable launch systems require maintenance, which is often substantial. The Space Shuttle system required extensive refurbishing between flights, primarily dealing with the silica tile TPS and the high performance LH2/LOX burning main engines. Both systems require a significant amount of detailed inspection, rebuilding and parts replacement between flights, and account for over 75% of the maintenance costs of the Shuttle system. These costs, far in excess of what had been anticipated when the system was constructed, have cut the maximum flight rate of Shuttle to 1/4 of that planned. This has also quadrupled the cost per pound of payload to orbit, making Shuttle economically infeasible in today's launch market for any but the largest payloads, for which there is no competition.

For any RLV technology to be successful, it must learn from the failings of Shuttle and overcome those failings with new technologies in the TPS and propulsion areas.

Manpower and Logistics

The Space Shuttle program required a standing army of over 9,000 employees to maintain, refurbish, and relaunch the shuttle fleet, irrespective of flight rates. That manpower budget must be divided by the total number of flights per year. The fewer flights means the cost per flight goes up significantly. Streamlining the manpower requirements of any launch system is an essential part of making an RLV economical. Projects that have attempted to develop this ethic include the DC-X Delta Clipper project, as well as SpaceX's Falcon 9 and Falcon 1 programs.

One issue mitigating against this drive for labor savings is government regulation. Given that NASA and USAF (as well as government programs in other countries) are the primary customers and sources of development capital, government regulatory requirements for oversight, parwork, quality, safety, and other documentation tend to inflate the operational costs of any such system.

Orbital Reusable Launchers

Under Development

- RLV-TD, (India): On 23 May 2016, ISRO successfully performed test flight of India's first reusable launch vehicle that operates at hypersonic speed.

- Adeline - Reusable launch system concept developed by Airbus Defence and Space

- Avatar RLV - Under development, first scaled-down demonstration flight planned in 2016.

- Blue Origin is developing a reusable booster system, as of November 2015. Blue Origin New Shepard rocket is the first rocket successfully launched and which is proven to be able to land vertically on earth VTVL after reaching space, by passing Kármán line.

- As of 2014, China is working on a project to recover rocket boosters, using a "paraglider-type wings" approach. Powered flight tests are in the future, and the process is expect to take until approximately 2018.

- Skylon (spacecraft) (proposed airbreathing SSTO spaceplane)

- SpaceX reusable rocket launching system—(currently in development and test)—is planned for use on both the Falcon 9 and Falcon Heavy launch vehicles. A second-generation VTVL reusable design was publicly announced in 2011. The low-altitude flight test program of an experimental technology-demonstrator launch vehicle began in 2012, with more extensive high-altitude over-water flight testing planned to begin in mid-2013, and continue on each subsequent Falcon 9 flight. On December 21, 2015, SpaceX successfully landed a Falcon 9 first stage after it boosted 11 commercial satellites into low earth orbit on Falcon 9 Flight 20.

- Swiss Space Systems is developing launching system including the suborbital spaceplane SOAR. The first 2 stages, an Airbus 300 and SOAR, are completely reusable.

- zero2infinity is developing a launching system called bloostar based on the rockoon system, which consists in elevating to the near space the launcher using a high-altitude balloon and once there launch a multi-stage rocket to put a satellite into orbit.

Proposed and Concept Vehicles

- SpaceFleet (The EARL Project - a space-capable UAV concept

- SpaceLiner (a mid-2000s German proposed suborbital, hypersonic, winged passenger transport concept)

- Shenlong (spacecraft) (an early 2000s Chinese proposed, scaled model tested at high altitude in 2005)

- PlanetSpace Silver Dart (a 2000s partially reusable spaceplane concept, based on a hypersonic glider design)

- As of January 2015, the French space agency CNES is working with Germany and a few other governments to start a modest research effort with a hope to propose a LOX/methane engine on a reusable launch vehicle by mid-2015, with flight testing unlikely before approximately 2026.

- Sura (uk) (proposed by Ukraine)

- CORONA (proposed by Russia)

Historical Developed

- Buran (partially reusable, retired)

- Space Shuttle (partially reusable, retired)

Cancelled

- Baikal French/Russian early-2000s joint-project concept. Cancelled after "CNES officials concluded that a rocket system with a reusable first stage would need to launch some 40 times a year" in order to make the project economically feasible.

- HOTOL British SSTO.

- Hyperion SSTO 1960s concept HTVL spacecraft.

- Kliper planned Russian partly reusable orbiter, cancelled in 2006.

- Liquid Fly-back Booster proposed design of reusable boosters for Ariane 5 with additional derivatives

- MAKS proposed Russian system of Buran-like smaller winged reusable orbiter on heavy aircraft carrier.

- Spiral cancelled Soviet military system of small winged reusable orbiter on winged hypersonic air-carrier.

- Phoenix SSTO

- X-30 NASP, X-33 and VentureStar proposed SSTO replacement for the Space Shuttle, cancelled in 2001.

- Roton Commercial launch vehicle project, cancelled in 2000 due to lack of funds.

Reusability dropped, Flown only as Expendable

- SpaceX Falcon 1 was announced as a partially reusable launch vehicle, and the 28 September 2008 test flight reached orbit, but vehicle recovery was never demonstrated and the vehicle was retired after 2009.

Regulations

In 2006, the US Federal Aviation Administration issued a new regulation regarding commercial reusable launch vehicles, both suborbital and orbital, as Part 431. The text can be found under the US Federal Code at 14 CFR Part 431. The new regulation was made in anticipation of planned commercial reusable launch operations including the American companies listed above. FAA regulations only have jurisdiction within the United States and its territories, and to aircraft and spacecraft registered in the United States.

References

- "Blue Origin Makes Historic Reusable Rocket Landing in Epic Test Flight". Calla Cofield. Space. Com. 2015-11-24. Retrieved 2015-11-25.

- Berger, Eric. "Jeff Bezos and Elon Musk spar over gravity of Blue Origin rocket landing". Ars Technica. Retrieved 25 November 2015.

- "India's Futuristic Unmanned Space Shuttle Getting Final Touches". EXPRESS NEWS SERVICE. Indian Defence Research Wing. 20 May 2015. Retrieved 2015-08-02.

- "Wednesday, August 03, 2011India's Space Shuttle [Reusable Launch Vehicle (RLV)] program". AA Me, IN. 2011. Retrieved 2015-08-02.

- Reyes, Tim (October 17, 2014). "Balloon launcher Zero2Infinity Sets Its Sights to the Stars". Universe Today. Retrieved 9 July 2015.

- de Selding, Peter B. (5 January 2015). "With Eye on SpaceX, CNES Begins Work on Reusable Rocket Stage". SpaceNews. Retrieved 6 January 2015.

- Henry, Caleb (October 16, 2014). "Zero2infiniti Announces Bloostar Launch Vehicle, More than $200 Million Pre-Booked Sales". Satellite Today. Retrieved 9 July 2015.

- "The Maglifter: An Advanced Concept Using Electromagnetic Propulsion in Reducing the Cost of Space Launch". NASA. Retrieved 24 May 2011.

Aircraft and its Types

Aircrafts are machines which are able to fly by the support of air. A monoplane has a single main wing place, in difference to a biplane. Zeppelin, monoplane, rotorcraft, helicopter and powered aircraft are some of the aircrafts which have been explained in the chapter. Aircrafts and helicopters are one of the significant and important topics related to aerospace engineering. The following chapter unfolds its crucial aspects in a critical yet systematic manner.

Zeppelin

A Zeppelin was a type of rigid airship named after the German Count Ferdinand von Zeppelin who pioneered rigid airship development at the beginning of the 20th century. Zeppelin's ideas were first formulated in 1874 and developed in detail in 1893. They were patented in Germany in 1895 and in the United States in 1899. After the outstanding success of the Zeppelin design, the word *zeppelin* came to be commonly used to refer to all rigid airships. Zeppelins were first flown commercially in 1910 by Deutsche Luftschiffahrts-AG (DELAG), the world's first airline in revenue service. By mid-1914, DELAG had carried over 10,000 fare-paying passengers on over 1,500 flights. During World War I the German military made extensive use of Zeppelins as bombers and scouts, killing over 500 people in bombing raids in Britain.

The USS *Los Angeles*, a US Navy airship built by the Zeppelin Company

The defeat of Germany in 1918 temporarily slowed down the airship business. Although DELAG established a scheduled daily service between Berlin, Munich, and Friedrichshafen in 1919, the airships built for this service eventually had to be surrendered under the terms of the Treaty of Versailles, which also prohibited Germany from building large airships. An exception was made allowing the construction of one airship for the US Navy, which saved the company from extinction. In 1926 the restrictions on airship construction were lifted and with the aid of donations from the public work was started on the construction of LZ 127 *Graf Zeppelin*. This revived the company's fortunes, and during the 1930s the airships *Graf Zeppelin* and the larger LZ 129 *Hindenburg* operated regular transatlantic flights from Germany to North America and Brazil. The Art Deco spire of the Empire State Building was originally designed to serve as a mooring mast for Zeppelins and other airships, although it was found that high winds made this impossible and the plan was abandoned. The *Hindenburg* disaster in 1937, along with political and economic issues, hastened the demise of the Zeppelins.

Principal Characteristics

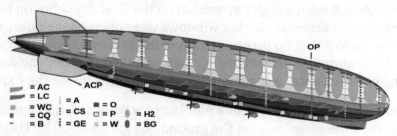

The pink ovals depict hydrogen cells inside the LZ 127, the magenta elements are *Blaugas* cells. The full-resolution picture labels more internals.

The principal feature of Zeppelin's design was a fabric-covered rigid metal framework made up from transverse rings and longitudinal girders containing a number of individual gasbags. The advantage of this design was that the aircraft could be much larger than non-rigid airships, which relied on a slight overpressure within the single pressure envelope to maintain their shape. The framework of most Zeppelins was made of duralumin (a combination of aluminum and copper as well as two or three other metals— its exact content was kept a secret for years). Early Zeppelins used rubberised cotton for the gasbags, but most later craft used goldbeater's skin, made from the intestines of cattle.

The first Zeppelins had long cylindrical hulls with tapered ends and complex multiplane fins. During World War I, following the lead of their rivals Schütte-Lanz Luftschiffbau, the design changed to the more familiar streamlined shape with cruciform tail surfaces, as used by almost all later airships.

They were propelled by several engines, mounted in gondolas or engine cars, which were attached to the outside of the structural framework. Some of these could provide reverse thrust for manoeuvring while mooring.

Early models had a comparatively small externally mounted gondola for passengers and crew which was attached to the bottom of the frame. This space was never heated (fire outside of the kitchen was considered too risky) so passengers during trips across the North Atlantic or Siberia were forced to bundle themselves in blankets and furs to keep warm and were often miserable with the cold.

Keel plan of the Hindenburg

By the time of the Hindenburg, several important changes had taken place: the passenger space had been relocated to the interior of the overall vessel, passenger rooms were insulated from the exterior by the dining area, and forced-warm air could be circulated from the water that cooled the forward engines, all of which made traveling much more comfortable though it deprived passengers of views from the windows of their berths which had been a major attraction on the Graf Zeppelin: on both the older and newer vessels, the external viewing windows were often opened during flight. The flight ceiling was so low that no pressurization of the cabins was necessary, though the Hindenburg did maintain a pressurized air-locked smoking room (no flame allowed, however— one electric lighter was maintained permanently inside the room).

Access to the Zeppelin was achieved in a number of ways. The Graf Zeppelin's gondola was accessed while the vessel was on the ground, via gangways. The Hindenburg also had passenger gangways that led from the ground directly into its hull and which could then be withdrawn entirely as well as ground access to the gondola and an exterior access hatch via its electrical room which was intended for crew use only.

History

Early Designs

Ferdinand von Zeppelin

Count Ferdinand von Zeppelin's serious interest in airship development began in 1874, when he took inspiration from a lecture given by Heinrich von Stephan on the subject of "World Postal Services and Air Travel" to outline the basic principle of his later craft in a diary entry dated 25 March 1874. This describes a large rigidly framed outer envelope containing several separate gasbags. He had previously encountered Union Army balloons in 1863 when he visited the United States as a military observer during the American Civil War.

Count Zeppelin began to seriously pursue his project after his early retirement from the military in 1890 at the age of 52. Convinced of the potential importance of aviation, he started working on various designs in 1891, and had completed detailed designs by 1893. An official committee reviewed his plans in 1894, and he received a patent, granted on 31 August 1895, with Theodor Kober producing the technical drawings.

Zeppelin's patent described a *Lenkbares Luftfahrzug mit mehreren hintereinanderen angeordneten Tragkörpern* [Steerable airship-train with several carrier structures arranged one behind another], - an airship consisting of flexibly articulated rigid sections. The front section, containing the crew and engines, was 117.35 m (385 ft) long with a gas capacity of 9514 cu m (336,000 cu ft): the middle section was 16 m (52 ft 6 in) long with an intended useful load of 599 kg (1320 lb) and the rear section 39.93 m (131 ft) long with an intended load of 1996 kg (4,400 lb)

Count Zeppelin's attempts to secure government funding for his project proved unsuccessful, but a lecture given to the Union of German Engineers gained their support. Zeppelin also sought support from the industrialist Carl Berg, then engaged in construction work on the second airship design of David Schwarz. Berg was under contract not to supply aluminium to any other airship manufacturer, and subsequently made a payment to Schwartz's widow as compensation for breaking this agreement. Schwarz's design differed fundamentally from Zeppelin's, crucially lacking the use of separate gasbags inside a rigid envelope,

The first flight of LZ 1 over Lake Constance (the *Bodensee*) in 1900

In 1898 Count Zeppelin founded the *Gesellschaft zur Förderung der Luftschiffahrt* (Society for the Promotion of Airship Flight), contributing more than half of its 800,000 mark share-capital himself. Responsibility for the detail design was given to Kober, whose place was later taken by Ludwig Dürr, and construction of the first airship began in 1899 in a floating assembly-hall in the Bay of Manzell near Friedrichshafen on Lake Constance (the *Bodensee*). The intention behind the floating hall was to facilitate the difficult task of bringing the airship out of the hall, as it could easily be aligned with the wind. The LZ 1 (LZ for *Luftschiff Zeppelin*, or "Zeppelin Airship") was 128 metres (420 ft) long with a hydrogen capacity of 11,000 m³ (400,000 cu ft), was driven by two 15 horsepower (11 kW) Daimler engines each driving a pair of propellers mounted either side of the envelope via bevel gears and a driveshaft, and was controlled in pitch by moving a weight between its two nacelles.

The first flight took place on 2 July 1900 over Lake Constance. Damaged during landing, it was repaired and modified and proved its potential in two subsequent flights made on 17 and 24 October 1900, bettering the 6 m/s (21.6 km/h, 13.4 mph) velocity attained by the French airship *La France*. Despite this performance, the shareholders declined to invest more money, and so the company was liquidated, with Count von Zeppelin purchasing the ship and equipment. The Count wished to continue experimenting, but he eventually dismantled the ship in 1901.

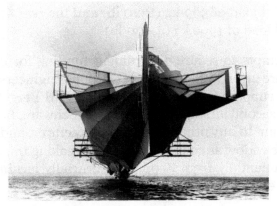

Zeppelin LZ 4 with its multiple stabilizers, 1908

Donations, the profits of a special lottery, some public funding, a mortgage of Count von Zeppelin's wife's estate and a 100,000 mark contribution by Count von Zeppelin himself allowed the construction of LZ 2, which made only a single flight on 17 January 1906. After both engines failed it made a forced landing in the Allgäu mountains, where a storm subsequently damaged the anchored ship beyond repair.

Incorporating all the usable parts of LZ 2, its successor LZ 3 became the first truly successful Zeppelin. This renewed the interest of the German military, but a condition of purchase of an airship was a 24-hour endurance trial. This was beyond the capabilities of LZ 3, leading Zeppelin to construct his fourth design, the LZ 4, first flown on 20 June 1908. On 1 July it was flown over Switzerland to Zürich and then back to Lake Constance, covering 386 km (240 mi) and reaching an altitude of 795 m (2,600 ft). An attempt to complete the 24-hour trial flight ended when LZ 4 had to make a landing at Echterdingen near Stuttgart because of mechanical problems. During the stop, a storm tore the airship away from its moorings on the afternoon of 5 August 1908. It crashed into a tree, caught fire, and quickly burnt out. No one was seriously injured.

Wreckage of LZ 4

This accident would have finished Zeppelin's experiments, but his flights had generated huge public interest and a sense of national pride regarding his work, and spontaneous donations from the public began pouring in, eventually totalling over six million marks. This enabled the Count to found the *Luftschiffbau Zeppelin GmbH* (Airship Construction Zeppelin Ltd.) and the Zeppelin Foundation.

Before World War I

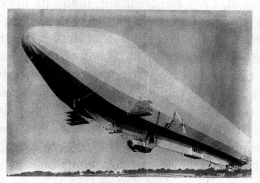

LZ 7 *Deutschland*

Before World War I (1914–1918) the Zeppelin company manufactured 21 more airships. The Imperial German Army bought LZ 3 and LZ 5 (a sister-ship to LZ 4 which was completed in May 1909) and designated them Z 1 and Z II respectively. Z II was wrecked in a gale in April 1910, while Z I flew until 1913, when it was decommissioned and replaced by LZ 15, designated *ersatz* Z I. First flown on 16 January 1913, it was wrecked on 19 March of the same year. In April 1913 its newly built sister-ship LZ 15 (Z IV) accidentally intruded into French airspace owing to a navigational error caused by high winds and poor visibility. The commander judged it proper to land the airship to demonstrate that the incursion was accidental, and brought the ship down on the military parade-ground at Lunéville. The airship remained on the ground until the following day, permitting a detailed examination by French airship experts.

In 1909 Count Zeppelin founded the world's first airline, the Deutsche Luftschiffahrts-Aktiengesellschaft (German Airship Travel Corporation), generally known as DELAG to promote his airships, initially using LZ 6, which he had hoped to sell to the German Army. The airships did not provide a scheduled service between cities, but generally operated pleasure cruises, carrying twenty passengers. The airships were given names in addition to their production numbers. LZ 6 first flew on 25 August 1909 and was accidentally destroyed in Baden-Oos on 14 September 1910 by a fire in its hangar.

A monument near Bad Iburg commemorating the 1910 LZ 7 crash

The second DELAG airship, LZ 7 *Deutschland*, made its maiden voyage on 19 June 1910. On 28 June it set off on a voyage to publicise Zeppelins, carrying 19 journalists as passengers. A combination of adverse weather and engine failure brought it down at Mount Limberg near Bad Iburg in Lower Saxony, its hull getting stuck in trees. All passengers and crew were unhurt, except for one crew member who broke his leg when he jumped from the craft. It was replaced by LZ 8 *Deutschland II* which also had a short career, first flying on 30 March 1911 and becoming damaged beyond repair when caught by a strong cross-wind when being walked out of its shed on 16 May. The company's fortunes changed with the next ship, LZ 10 *Schwaben*, which first flew on 26 June 1911 and carried 1,553 passengers in 218 flights before catching fire after a gust tore it from its mooring near Düsseldorf. Other DELAG ships included LZ 11 *Viktoria Luise* (1912), LZ 13 *Hansa* (1912) and *LZ 17* and LZ 17*Sachsen* (1913). By the outbreak of World War I in August 1914 1588 flights had carried 10,197 fare-paying passengers.

LZ 18 (L 2)

On 24 April 1912 the Imperial German Navy ordered its first Zeppelin - an enlarged version of the airships operated by DELAG - which received the naval designation Z 1 and entered Navy service in October 1912. On 18 January 1913 Admiral Alfred von Tirpitz, Secretary of State of the German Imperial Naval Office, obtained the agreement of Kaiser Wilhelm II to a five-year program of expansion of German naval-airship strength, involving the building of two airship bases and constructing a fleet of ten airships. The first airship of the program. L 2, was ordered on 30 January. L 1 was lost on 9 September near Heligoland when caught in a storm while taking part in an exercise with the German fleet. 14 crew members drowned, the first fatalities in a Zeppelin accident. Less than six weeks later, on 17 October, LZ 18 (L 2) caught fire during its acceptance trials, killing the entire crew. These accidents deprived the Navy of most of its experienced personnel: the head of the Admiralty Air Department was killed in the L 1 and his successor died in the L 2. The Navy was left with three partially trained crews. The next Navy zeppelin, the M class L 3 did not enter service until May 1914: in the meantime, *Sachsen* was hired from DELAG as a training ship.

By the outbreak of war in August 1914 Zeppelin had started constructing the first M class airships, which had a length of 158 m (518 ft), with a volume of 22,500 cubic metres (794,500 cu ft) and a useful load of 9,100 kilograms (20,100 lb). Their three Maybach C-X engines produced a 470 kilowatts (630 hp) each, and they could reach speeds of up to 84 kilometres per hour (52 mph).

During World War I

The German airships were operated by the Army and Navy as two entirely separate organizations.

Crater of a Zeppelin bomb in Paris, 1916

When World War I broke out, the Army took over the three remaining DELAG ships. By this time, it had already decommissioned three older Zeppelins, including Z I. During the war the Navy Zeppelins were mainly used in reconnaissance missions . Bombing missions, especially those targeting London, captured the German public's imagination, but had little significant material success, although the Zeppelin raids, together with the later bombing raids carried out by aeroplanes, did cause the diversion of men and equipment from the Western Front and fear of the raids had some effect on industrial production.

Early offensive operations by Army airships revealed that they were extremely vulnerable to ground fire unless flown at high altitude, and several were lost. No bombs had been developed, and the early raids dropped artillery shells instead. On 5 August 1914 Z VI bombed Liège. Flying at a relatively low altitude because of cloud cover, the airship was damaged by small-arms fire and was destroyed in a forced landing near Bonn. On 21 August Z VII and Z VIII were damaged by ground fire while supporting German army operations in Alsace, and Z VIII was lost. On the night of 24/25 August Z IX bombed Antwerp, dropping bombs near the royal palace and killing five people. A second, less effective, raid was made on the night of 1–2 September and a third on 7 October, but on 8 October Z IX was destroyed in its hangar at Düsseldorf by Flt Lt. Reginald Marix, RNAS. The RNAS had also bombed the Zeppelin bases in Cologne on 22 September 1914. On the eastern front, Z V was brought down by ground fire on 28 August during the Battle of Tannenberg; most of the crew were captured. Z IV bombed Warsaw on 24 September and was also used to support German army operations in East Prussia. By the end of 1914 the Army's airship strength was reduced to four.

On 20 March 1915, temporarily forbidden from bombing London by the Kaiser, Z X (LZ 29), LZ 35 and the Schütte-Lanz airship SL 2 set off to bomb Paris: SL 2 was damaged by artillery fire while crossing the front and turned back but the two Zeppelins reached Paris and dropped 1,800 kg (4,000 lb) of bombs, killing only one and wounding eight. On the return journey Z X was damaged by anti-aircraft fire and was damaged beyond repair in the resulting forced landing. Three weeks later LZ 35 suffered a similar fate after bombing Poperinghe. Two further missions were flown against Paris in January 1916: on 29 January LZ 79 killed 23 and injured another 30 but was so severely damaged by anti-aircraft fire that it crashed during the return journey. A second mission by LZ 77 the following night bombed the suburbs of Asnières and Versailles, with little effect.

Airship operations in the Balkans started in the autumn of 1915, and an airship base was constructed at Szentandras. In November 1915 LZ 81 was used to fly diplomats to Sofia for negotiations with the Bulgarian government. This base was also used by LZ 85 to conduct two raids on Salonika in early 1916: a third raid on 4 May ended with it being brought down by antiaircraft fire. The crew survived but were taken prisoner. When Romania entered the war in August 1916 LZ 101 was transferred to Yambol and bombed Bucharest on 28 August, 4 September and 25 September. LZ 86, transferred to Szentandras and made a single attack on the Ploieşti oil fields on 4 September but was wrecked on attempting to land after the mission. Its replacement, LZ 86, was damaged by antiaircraft fire during its second attack on Bucharest on 26 September and was damaged beyond repair in the resulting forced landing, and was replaced by LZ 97.

At the instigation of the Kaiser a plan was made to bomb St Petersburg in December 1916. Two Navy zeppelins were transferred to Wainoden on the Courland Peninsula. A preliminary attempt to bomb Reval on 28 December ended in failure caused by operating problems due to the extreme cold, and one of the airships was destroyed in a forced landing at Serappen. The plan was subsequently abandoned.

Wreckage of Zeppelin L31 or L32 shot down over England 23 Sept 1916.

In 1917 the German High Command made an attempt to use a Zeppelin to deliver supplies to Lettow-Vorbeck's forces in German East Africa. L 57, a specially lengthened craft was to have flown the mission but was destroyed shortly after completion. A Zeppelin then under construction, L 59, was then modified for the mission: it set off from Yambol on 21 November 1917 and nearly reached its destination, but was ordered to return by radio. Its journey covered 6,400 km (4,000 mi) and lasted 95 hours. It was then used for reconnaissance and bombing missions in the eastern Mediterranean. It flew one bombing mission against Naples on 10–11 March 1918. A planned attack on Suez was turned back by high winds, and on 7 April 1918 it was on a mission to bomb the British naval base at Malta when it caught fire over the Straits of Otranto, with the loss of all its crew.

On 5 January 1918, a fire at Ahlhorn destroyed four of the specialised double sheds along with four Zeppelins and one Schütte-Lanz. In July 1918, the Tondern Raid conducted by the RAF and Royal Navy, destroyed two Zeppelins in their sheds.

1914–18 Naval Patrols

A Zeppelin flying over SMS *Seydlitz*

The main use of the airship was in reconnaissance over the North Sea and the Baltic, and the majority of airships manufactured were used by the Navy. Patrolling had priority over any other airship activity. During the war almost 1,000 missions were flown over the North Sea alone, compared to about 50 strategic bombing raids. The German Navy had some 15 Zeppelins in commission by the end of 1915 and was able to have two or more patrolling continuously at any one time. However their operations were limited by weather conditions. On 16 February L 3 and L 4 were lost owing to a combination of engine failure and high winds, L 3 crashing on the Danish island of Fanø without loss of life and L 4 coming down at Blaavands Huk; eleven crew escaped from the forward gondola, after which the lightened airship with four crew members remaining in the aft engine car was blown out to sea and lost.

At this stage in the war there was no clear doctrine for the use of Naval airships. A single Zeppelin, L 5, played an unimportant part in the Battle of the Dogger Bank on 24 January 1915. L 5 was carrying out a routine patrol when it picked up Admiral Hipper's radio signal announcing that he was engaged with the British battle cruiser squadron. Heading towards the German fleet's position, the Zeppelin was forced to climb above the cloud cover by fire from the British fleet: its commander then decided that it was his duty to cover the retreating German fleet rather than observe British movements.

In 1915 patrols were only carried out on 124 days, and in other years the total was considerably less. They prevented British ships from approaching Germany, spotted when and where the British were laying mines and later aided in the destruction of those mines. Zeppelins would sometimes land on the sea next to a minesweeper, bring aboard an officer and show him the mines' locations.

In 1917 the British Navy began to take effective countermeasures against airship patrols over the North Sea. In April the first Curtiss H.12 "Large America" long-range flying boats were delivered to RNAS Felixstowe, and in July 1917 the aircraft carrier HMS *Furious* entered service, and launching platforms for aeroplanes were fitted to the forward turrets of some light cruisers. On 14 May L 22 was shot down near Terschelling Bank by an H.12 flown by Lt. Galpin and Sub-Lt. Leckie which had been alerted following interception of its radio traffic. Two further unsuccessful interceptions were made by Galpin and Leckie on 24 May and 5 June, and on 14 June L 43 was brought down by an H.12 flown by Sub Lts. Hobbs and Dickie. On the same day Galpin and Leckie intercepted and attacked L 46. The Germans had believed that the previous unsuccessful attacks had been made by an aircraft operating from one of the British Navy's seaplane carriers: now realising that there was a new threat, Strasser ordered airships patrolling in the Terschilling area to maintain an altitude of at least 4,000 m (13,000 ft), considerably reducing their effectiveness. On 21 August L 23, patrolling off the Danish coast, was spotted by the British 3rd Light Cruiser squadron which was in the area. HMS *Yarmouth* launched its Sopwith Pup, and Sub-Lt. B. A. Smart succeeded in shooting the Zeppelin down in flames. The cause of the airship's loss was not discovered by the Germans, who believed the Zeppelin had been brought down by antiaircraft fire from surface ships.

Bombing Campaign Against Britain

At the beginning of the conflict the German command had high hopes for the airships, which were considerably more capable than contemporary light fixed-wing machines: they were almost as fast, could carry multiple machine guns, and had enormously greater bomb-load range and endurance. Contrary to expectation, it was not easy to ignite the hydrogen using standard bullets and shrapnel. The Allies only started to exploit the Zeppelin's great vulnerability to fire when a combination of explosive and incendiary ammunition was introduced during 1916. The British had been concerned over the threat posed by Zeppelins since 1909, and attacked the Zeppelin bases early in the war. LZ 25 was destroyed in its hangar at Düsseldorf on 8 October 1914 by bombs dropped by Flt Lt Reginald Marix, RNAS, and the sheds at Cologne as well as the Zeppelin works in Friedrichshafen were also attacked. These raids were followed by the Cuxhaven Raid on Christmas Day 1914, one of the first operations carried out by ship-launched aeroplanes.

British First World War poster of a Zeppelin above London at night

Airship raids on Great Britain were approved by the Kaiser on 7 January 1915, although he excluded London as a target and further demanded that no attacks be made on historic buildings. The raids were intended to target only military sites on the east coast and around the Thames estuary, but bombing accuracy was poor owing to the height at which the airships flew and navigation was problematic. The airships relied largely on dead reckoning, supplemented by a radio direction-finding system of limited accuracy. After blackouts became widespread, many bombs fell at random on uninhabited countryside.

1915

The first raid on England took place on the night of 19–20 January 1915. Two Zeppelins, L 3 and L 4, intended to attack Humberside but, diverted by strong winds, eventually dropped their bombs on Great Yarmouth, Sheringham, King's Lynn and the surrounding villages, killing four and injuring 16. Material damage was estimated at £7,740.

The Kaiser authorised the bombing of the London docks on 12 February 1915, but no raids on London took place until May. Two Navy raids failed due to bad weather on 14 and 15 April, and it was decided to delay further attempts until the more capable P class Zeppelins were in service. The Army received the first of these, LZ 38, and Erich Linnarz commanded it on a raid over Ipswich on 29–30 April and another, attacking Southendon 9–10 May. LZ 38 also attacked Dover and Ramsgate on 16–17 May, before returning to bomb Southend on 26–27 May. These four raids killed six people and injured six, causing property damage estimated at £16,898. Twice Royal Naval Air Service (RNAS) aircraft tried to intercept LZ 38 but on both occasions it was either able to outclimb the aircraft or was already at too great an altitude for the aircraft to intercept.

On 31 May Linnarz commanded LZ 38 on the first raid against London. In total some 120 bombs were dropped on a line stretching from Stoke Newington south to Stepney and then north toward Leytonstone. Seven people were killed and 35 injured. 41 fires were started, burning out seven buildings and the total damage was assessed at £18,596. Aware of the problems that the Germans were experiencing in navigation, this raid caused the government to issue a D notice prohibiting the press from reporting anything about raids that was not mentioned in official statements. Only one of the 15 defensive sorties managed to make visual contact with the enemy, and one of the pilots, Flt Lieut D. M. Barnes, was killed on attempting to land.

The first naval attempt on London took place on 4 June: strong winds caused the commander of L 9 to misjudge his position, and the bombs were dropped on Gravesend. L 9 was also diverted by the weather on 6–7 June, attacking Hull instead of London and causing considerable damage. On the same night an Army raid of three Zeppelins also failed because of the weather, and as the airships returned to Evére they ran into a counter-raid by RNAS aircraft flying from Furnes, Belgium. LZ 38 was destroyed on the ground and LZ 37 was intercepted in the air by R. A. J. Warneford, who dropped six bombs on the airship, setting it on fire. All but one of the crew died. Warneford was awarded the Victoria Cross for his achievement. As a consequence of the RNAS raid both the Army and Navy withdrew from their bases in Belgium.

After an ineffective attack by L 10 on Tyneside on 15–16 June the short summer nights discouraged further raids for some months, and the remaining Army Zeppelins were reassigned to the Eastern and Balkan fronts. The Navy resumed raids on Britain in August, when three largely ineffective raids were carried out. On 10 August the antiaircraft guns had their first success, causing L 12 to come down into the sea off Zeebrugge, and on 17–18 August L 10 became the first Navy airship to reach London. Mistaking the reservoirs of the Lea Valley for the Thames, it dropped its bombs on Walthamstow and Leytonstone. L 10 was destroyed a little over two weeks later: it was struck by lightning and caught fire off Cuxhaven, and the entire crew was killed. Three Army airships set off to bomb London on 7–8 September, of which two succeeded: SL 2 dropped bombs between Southwark and Woolwich: LZ 74 scattered 39 bombs over Cheshunt before heading on to London and dropping a single bomb on Fenchurch Street station.

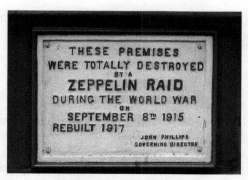

A commemorative plaque at 61 Farringdon Road, London

The Navy attempted to follow up the Army's success the following night. One Zeppelin targeted the benzol plant at Skinningrove and three set off to bomb London: two were forced to turn back but L 13, commanded by *Kapitänleutnant* Heinrich Mathy reached London. The bomb-load included a 300 kilograms (660 lb) bomb, the largest yet carried. This exploded near Smithfield Market, destroying several houses and killing two men. More bombs fell on the textile warehouses north of St Paul's Cathedral, causing a fire which despite the attendance of 22 fire engines caused over half a million pounds of damage: Mathy then turned east, dropping his remaining bombs on Liverpool Street station. The Zeppelin was the target of concentrated antiaircraft fire, but no hits were scored and the falling shrapnel caused both damage and alarm on the ground. The raid killed 22 people and injured 87. The monetary damage was over one sixth of the total damage inflicted by bombing raids during the war.

After three more raids were scattered by the weather, a five-Zeppelin raid was launched by the Navy on 13 October, the "Theatreland Raid." Arriving over the Norfolk coast at around 18:30, the Zeppelins encountered new ground defences installed since the September raid; these had no success, although the airship commanders commented on the improved defences of the city. L 15 began bombing over Charing Cross, the first bombs striking the Lyceum Theatre and the corner of Exeter and Wellington Streets, killing 17 and injuring 20. None of the other Zeppelins reached central London: bombs fell on Woolwich, Guildford, Tonbridge, Croydon, Hertford and an army camp near Folkestone. A total of 71 people were killed and 128 injured. This was the last raid of 1915, as bad weather coincided with the new moon in both November and December 1915 and continued into January 1916.

Although these raids had no significant military impact, the psychological effect was considerable. The writer D. H. Lawrence described one raid in a letter to Lady Ottoline Morrell:

Then we saw the Zeppelin above us, just ahead, amid a gleaming of clouds: high up, like a bright golden finger, quite small (...) Then there was flashes near the ground — and the shaking noise. It was like Milton — then there was war in heaven. (...) I cannot get over it, that the moon is not Queen of the sky by night, and the stars the lesser lights. It seems the Zeppelin is in the zenith of the night, golden like a moon, having taken control of the sky; and the bursting shells are the lesser lights.

1916

The raids continued in 1916. In December 1915 additional P class Zeppelins and the first of the new Q class airships, were delivered. The Q class was an enlargement of the P class with improved ceiling and bomb-load.

The Army took full control of ground defences in February 1916, and a variety of sub 4-inch (less than 102 mm) calibre guns were converted to anti-aircraft use. Searchlights were introduced, initially manned by police. By mid-1916, there were 271 anti-aircraft guns and 258 searchlights across England. Aerial defences against Zeppelins were divided between the RNAS and the Royal Flying Corps (RFC), with the Navy engaging enemy airships approaching the coast while the RFC took responsibility once the enemy had crossed the coastline. Initially the War Office had believed that the Zeppelins used a layer of inert gas to protect themselves from incendiary bullets, and favoured the use of bombs or devices like the Ranken dart. However, by mid-1916 an effective mixture of explosive, tracer and incendiary rounds had been developed. There were 23 airship raids in 1916, in which 125 tons of bombs were dropped, killing 293 people and injuring 691.

Zeppelin flagstone, Edinburgh

Zeppelin bomb, on display at the National Museum of Flight

Section of girder from Zeppelin shot down in England in 1916. Now at NPL

The first raid of 1916 was carried out by the German Navy. Nine Zeppelins were sent to Liverpool on the night of 31 January–1 February. A combination of poor weather and mechanical problems scattered them across the Midlands and several towns were bombed. A total of 61 people were reported killed and 101 injured by the raid. Despite ground fog, 22 aircraft took off to find the Zeppelins but none succeeded, and two pilots were killed when attempting to land. One airship, the L 19, came down in the North Sea because of engine failure and damage from Dutch ground–fire. Although the wreck stayed afloat for a while and was sighted by a British trawler, the boat's crew refused to rescue the Zeppelin crew because they were outnumbered, and all 16 crew died.

Further raids were delayed by an extended period of poor weather and also by the withdrawal of the majority of Naval Zeppelins in an attempt to resolve the recurrent engine failures. Three Zeppelins set off to bomb Rosyth on 5–6 March but were forced by high winds to divert to Hull, killing 18, injuring 52 and causing £25,005 damage. At the beginning of April raids were attempted on five successive nights. Ten airships set off on 31 March: most turned back and L 15, damaged by antiaircraft fire and an aircraft attacking using Ranken darts, came down in the sea near Margate. Most of the 48 killed in the raid were victims of a single bomb which fell on an Army billet in Cleethorpes. The following night two Navy Zeppelins bombed targets in the north of England, killing 22 and injuring 130. On the night of 2/3 April a six-airship raid was made, targeting the naval base at Rosyth, the Forth Bridge and London. None of the airships bombed their intended targets; 13 were killed, 24 injured and much of the £77,113 damage was caused by the destruction of a warehouse in Leith containing whisky. Raids on 4/5 April and 5/6 April had little effect, as did a five-Zeppelin raid on 25/6 April and a raid by a single Army Zeppelin the following night. On 2/3 July a nine-Zeppelin raid against Manchester and Rosyth was largely ineffective due to weather conditions, and one was forced to land in neutral Denmark, its crew being interned.

On 28–29 July the first raid to include one of the new and much larger R-class Zeppelins, L 31, took place. The 10-Zeppelin raid achieved very little; four turned back early and the rest wandered over a fog-covered landscape before giving up. Adverse weather dispersed raids on 30–31 July and 2–3 August, and on 8–9 August nine airships attacked Hull with little effect. On 24–25 August 12 Navy Zeppelins were launched: eight turned back without attacking and only Heinrich Mathy's L 31 reached London; flying above low clouds, 36 bombs were dropped in 10 minutes on south east London. Nine people were killed, 40 injured and £130,203 of damage was caused.

Zeppelins were very difficult to attack successfully at high altitude, although this also made accurate bombing impossible. Aeroplanes struggled to reach a typical altitude of 10,000 feet (3,000 m), and firing the solid bullets usually used by aircraft Lewis guns was ineffectual: they made small holes causing inconsequential gas leaks. Britain developed new bullets, the Brock containing inflammable potassium chlorate, and the Buckingham filled with phosphorus, to ignite the potassium chlorate and hence the Zeppelin's hydrogen. These had become available by September 1916.

The biggest raid to date was launched on 2–3 September, when twelve German Navy and four Army airships set out to bomb London. A combination of rain and snowstorms scattered the airships while they were still over the North Sea. Only one of the naval airships came within seven miles of central London, and both damage and casualties were slight. The newly commissioned Schütte-Lanz SL 11 dropped a few bombs on Hertfordshire while approaching London: it was picked up by searchlights as it bombed Ponders End and at around 02:15 it was intercepted by a BE2c flown by Lt. William Leefe Robinson, who fired three 40-round drums of Brocks and Buckingham ammunition into the airship. The third drum started a fire and the airship was quickly enveloped in flames. It fell to the ground near Cuffley, witnessed by the crews of several of the other Zeppelins and many on the ground; there were no survivors. The victory earned Leefe Robinson a Victoria Cross; the pieces of SL 11 were gathered up and sold as souvenirs by the Red Cross to raise money for wounded soldiers.

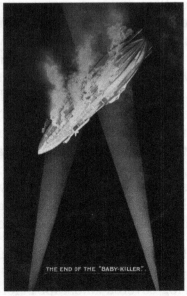

British propaganda postcard, entitled "The End of the 'Baby-Killer'"

The loss of SL 11 to the new ammunition ended the German Army's enthusiasm for raids on Britain. The German Navy remained aggressive, and another 12-Zeppelin raid was launched on 23–24 September. Eight older airships bombed targets in the Midlands and northeast, while four R-class Zeppelins attacked London. L 30 did not even cross the coast, dropping its bombs at sea. L 31 approached London from the south, dropping a few bombs on the southern suburbs before crossing the Thames and bombing Leyton, killing eight people and injuring 30. L 32 also approached from the south: it dropped a few bombs on Sevenoaks and Swanley before crossing Purfleet at about 01:00. Shortly afterwards it was found by a BE2c piloted by 2nd Lieutenant Frederick Sowrey and set alight, coming down near Great Burstead. The entire crew was killed. L 33 dropped a few incendiaries over Upminster and Bromley-by-Bow, where it was hit by an anti-aircraft shell, despite being at an altitude of 13,000 feet (4,000 m). As it headed towards Chelmsford it began to lose height and came down close to Little Wigborough. The airship was set alight by its crew, but inspection of the wreckage provided the British with much information about the construction of Zeppelins, which was used in the design of the British R33-class airships.

The next raid came on 1 October 1916. Eleven Zeppelins were launched at targets in the Midlands and at London. Only L 31, commanded by the experienced Heinrich Mathy making his 15th raid, reached London. As the airship neared Cheshunt at about 23:20 it was picked up by searchlights and attacked by three aircraft from No. 39 Squadron. 2nd lieutenant Wulstan Tempest succeeded in setting fire to the airship, which came down near Potters Bar. All 19 crew died, many jumping from the burning airship.

For the next raid, on 27–28 November, the Zeppelins avoided London for targets in the Midlands. Again the defending aircraft were successful: L 34 was shot down over the mouth of the Tees and L 21 was attacked by two aircraft and crashed into the sea off Lowestoft. There were no further raids in 1916 although the Navy lost three more craft, all on 28 December: SL 12 was destroyed at Ahlhorn by strong winds after sustaining damage in a poor landing, and at Tondern L 24 crashed into the shed while landing: the resulting fire destroyed both L 24 and the adjacent L 17.

1917

1917 watercolour by Felix Schwormstädt – translated title: "In the rear engine gondola of a Zeppelin airship during the flight through enemy airspace after a successful attack on England"

To counter the increasingly effective defences new Zeppelins were introduced with an increased operating altitude of 16,500 feet (5,000 m) and a ceiling of 21,000 feet (6,400 m). The first of these S-class Zeppelins, LZ 91 (L 42) entered service in February 1917. They were basically a modification of the R-class, sacrificing strength and power for improved altitude. The surviving R-class Zeppelins were adapted by removing one of the engines. The improved safety was offset by the extra strain on the airship crews caused by altitude sickness and exposure to extreme cold and operating difficulties caused by cold and unpredictable high winds encountered at altitude.

The first raid of 1917 did not occur until 16–17 March: the five high flying Zeppelins encountered very strong winds and none reached their targets. This experience was repeated on 23–24 May. Two days later 21 Gotha bombers attempted a daylight raid on London. They were frustrated by heavy cloud but the effort led the Kaiser to announce that airship raids on London were to stop; under pressure he later relented to allow the Zeppelins to attack under "favourable circumstances".

On 16–17 June, another raid was attempted. Six Zeppelins were to take part, but two were kept in their shed by high winds and another two were forced to return by engine failure. L 42 bombed Ramsgate, hitting a munitions store. The month-old L 48, the first U class Zeppelin, was forced to drop to 13,000 feet (4,000 m) where it was caught by four aircraft and destroyed, crashing near Theberton, Suffolk.

After ineffective raids on the Midlands and other targets in the north of England on 21–22 August and 24–25 September, the last major Zeppelin raid of the war was launched on 19–20 October, with 13 airships heading for Sheffield, Manchester and Liverpool. All were hindered by an unexpected strong headwind at altitude. L 45 was trying to reach Sheffield, but instead it dropped bombs on Northampton and London: most fell in the north-west suburbs but three 300 kg (660 lb) bombs fell in Piccadilly, Camberwell and Hither Green, causing most of the casualties that night. L 45 then reduced altitude to try to escape the winds but was forced back into the higher air currents by a BE2e. The airship then had mechanical failure in three engines and was blown over France, eventually coming down near Sisteron; it was set on fire and the crew surrendered. L 44 was brought down by ground fire over France: L 49 and L 50 were also lost to engine failure and the weather over France. L 55 was badly damaged on landing and later scrapped.

There were no more raids in 1917, although the airships were not abandoned but refitted with new, more powerful engines.

1918

There were only four raids in 1918, all against targets in the Midlands and northern England. Five Zeppelins attempted to bomb the Midlands on 12–3 March to little effect. The following night three Zeppelins set off, but two turned back because of the weather: the third bombed Hartlepool, killing eight and injuring 29. A five-Zeppelin raid on 12–13 April was also largely ineffective, with thick clouds making accurate navigation impossible. However some alarm was caused by the other two, one of which reached the east coast and bombed Wigan, believing it was Sheffield: the other bombed Coventry in the belief that it was Birmingham. The final raid on 5 August 1918 involved four airships and resulted in the loss of *L.70* and the death of its entire crew under the command of *Fregattenkapitän* Peter Strasser, head of the Imperial German Naval Airship Service and the *Führer der Luftschiffe*. Crossing the North Sea during daylight, the airship was intercepted by a Royal Air Force DH.4 biplane piloted by Major Egbert Cadbury, and shot down in flames.

Technological Progress

Zeppelin technology improved considerably as a result of the increasing demands of warfare. The company came under government control, and new personnel were recruited to the company to cope with the increased demand including the aerodynamacist Paul Jaray and the stress engineer Karl Arnstein. Many of these technological advances originated from Zeppelin's only serious competitor, the Mannheim-based Schütte-Lanz company. While their dirigibles were never as successful, Professor Schütte's more scientific approach to airship design led to important innovations including the streamlined hull shape, the simpler cruciform fins (replacing the more complicated box-like arrangements of older Zeppelins), individual direct-drive engine cars anti-aircraft machine-gun positions, and gas ventilation shafts which transferred vented hydrogen to the top of the airship. New production facilities were set up, assembling Zeppelins from components fabricated in Friedrichshafen.

The pre-war M-class designs were quickly enlarged, to produce the 163 metres (536 ft) long duralumin P-class, which increased gas capacity from 22,500 m³ (794,500 cu ft) to 31,900 m³ (1,126,000 cu ft), introduced a fully enclosed gondola and an extra engine. These modifications added 610 m (2,000 ft) to the maximum ceiling, around 9 km/h (6 mph) to the top speed, and greatly increased crew comfort and hence endurance. Twenty-two P-class airships were built; the first, LZ 38, was delivered to the Army on 3 April 1915. The P class was followed by a lengthened version, the Q class.

In July 1916 Luftschiffbau Zeppelin introduced the R-class, 199.49 m (644 ft 8 in) long, and with a volume of 55,210 m³ (1,949,600 cu ft). These could carry loads of three to four tons of bombs and reach speeds of up to 103 km/h (64 mph), powered by six 240 hp (180 kW) Maybach engines.

In 1917, following losses to air defences over Britain, new designs were produced which were capable of flying at much higher altitudes, typically operating at around 6,100 m (20,000 ft). This was achieved by reducing the weight of the airship by reducing the weight of the structure, halving the bomb load, deleting the defensive armament and by reducing the number of engines to five. However these were not successful as bombers: the greater height at which they operated greatly hindered navigation, and their reduced power made them vulnerable to unfavorable weather conditions.

The observation car preserved at the Imperial War Museum

At the beginning of the war Captain Ernst A. Lehmann and Baron Gemmingen, Count Zeppelin's nephew, developed an observation car for use by dirigibles. This was equipped with a wicker chair, chart table, electric lamp and compass, with telephone line and lightning conductor part of the suspension cable. The car's observer would relay navigation and bomb dropping orders to the Zeppelin flying within or above the clouds, so remaining invisible from the ground. Although used by Army airships, they were not used by the Navy, since Strasser considered that their weight meant an unacceptable reduction in bomb load.

End of the War

The German defeat also marked the end of German military dirigibles, as the victorious Allies demanded a complete abolition of German air forces and surrender of the remaining airships as reparations. Specifically, the Treaty of Versailles contained the following articles dealing explicitly with dirigibles:

Article 198

> "The armed forces of Germany must not include any military or naval air forces ... No dirigible shall be kept."

Article 202

> "On the coming into force of the present Treaty, all military and naval aeronautical material ... must be delivered to the Governments of the Principal Allied and Associated Powers ... In particular, this material will include all items under the following heads which are or have been in use or were designed for warlike purposes:

[...]

- "Dirigibles able to take to the air, being manufactured, repaired or assembled."

- "Plant for the manufacture of hydrogen."

- "Dirigible sheds and shelters of every kind for aircraft."

"Pending their delivery, dirigibles will, at the expense of Germany, be maintained inflated with hydrogen; the plant for the manufacture of hydrogen, as well as the sheds for dirigibles may at the discretion of the said Powers, be left to Germany until the time when the dirigibles are handed over."

On 23 June 1919, a week before the treaty was signed, many Zeppelin crews destroyed their airships in their halls in order to prevent delivery, following the example of the German fleet which had been scuttled two days before in Scapa Flow. The remaining dirigibles were transferred to France, Italy, Britain, and Belgium in 1920.

A total of 84 Zeppelins were built during the war. Over 60 were lost, roughly evenly divided between accident and enemy action. 51 raids had been made on England alone,[N 1] in which 5,806 bombs were dropped, killing 557 people and injuring 1,358 while causing damage estimated at £1.5 million. It has been argued the raids were effective far beyond material damage in diverting and hampering wartime production: one estimate is that the due to the 1915–16 raids "one sixth of the total normal output of munitions was entirely lost."

After World War I

Renaissance

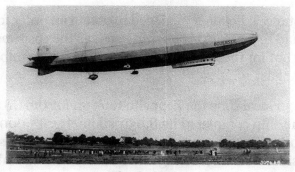

The *Bodensee*

Count von Zeppelin had died in 1917, before the end of the war. Dr. Hugo Eckener, who had long envisioned dirigibles as vessels of peace rather than of war, took command of the Zeppelin business, hoping to quickly resume civilian flights. Despite considerable difficulties, they completed two small passenger airships; LZ 120 *Bodensee*, which first flew in August 1919 and in the following months transported passengers between Frie-drichhafen and Berlin, and a sister-ship *LZ 121 Nordstern*, which was intended for use on a regular route to Stockholm.

However, in 1921 the Allied Powers demanded that these should be handed over as war reparations as compensation for the dirigibles destroyed by their crews in 1919. Germany was not allowed to construct military aircraft and only airships of less than 28,000 m³ (1,000,000 cu ft) were permitted. This brought a halt to Zeppelin's plans for airship development, and the company temporarily had to resort to manufacturing aluminium cooking utensils.

Eckener and his co-workers refused to give up and kept looking for investors and a way to circumvent Allied restrictions. Their opportunity came in 1924. The United States had started to experiment with rigid airships, constructing one of their own, the ZR-1 USS *Shenandoah*, and buying the R38 (based on the Zeppelin L 70) when the British airship programme was cancelled. However, this broke apart and caught fire during a test flight above the Humber on 23 August 1921, killing 44 crewmen.

ZR-3 USS *Los Angeles* over southern Manhattan

Under these circumstances, Eckener managed to obtain an order for the next American dirigible. Germany had to pay for this airship itself, as the cost was set against the war reparation accounts, but for the Zeppelin company this was unimportant. LZ 126 made its first flight on 27 August 1924.

On 12 October at 07:30 local time the Zeppelin took off for the US under the command of Hugo Eckener. The ship completed its 8,050 kilometres (5,000 mi) voyage without any difficulties in 80 hours 45 minutes. American crowds enthusiastically celebrated the arrival, and President Calvin Coolidge invited Eckener and his crew to the White House, calling the new Zeppelin an "angel of peace".

Given the designation ZR-3 USS *Los Angeles* and refilled with helium (partly sourced from the *Shenandoah*) after its Atlantic crossing, the airship became the most success-ful American airship. It operated reliably for eight years until it was retired in 1932 for economic reasons. It was dismantled in August 1940.

Golden Age

Graf Zeppelin under construction

With the delivery of *LZ 126*, the Zeppelin company had reasserted its lead in rigid airship construction, but it was not yet quite back in business. In 1926 restrictions on airship construction were relaxed by the Locarno treaties, but acquiring the necessary funds for the next project proved a problem in the difficult economic situation of post-World-War-I Germany, and it took Eckener two years of lobbying and publicity work to secure the realization of *LZ 127*.

Another two years passed before 18 September 1928, when the new dirigible, christened *Graf Zeppelin* in honour of the Count, flew for the first time. With a total length of 236.6 metres (776 ft) and a volume of 105,000 m³, it was the largest dirigible to have been built at the time. Eckener's initial purpose was to use *Graf Zeppelin* for experimental and demonstration purposes to prepare the way for regular airship traveling, carrying passengers and mail to cover the costs. In October 1928 its first long-range voyage brought it to Lakehurst, the voyage taking 112 hours and setting a new endurance record for airships. Eckener and his crew, which included his son Hans, were once more welcomed enthusiastically, with confetti parades in New York and another invitation to the White House. *Graf Zeppelin* toured Germany and visited Italy, Palestine, and Spain. A second trip to the United States was aborted in France due to engine failure in May 1929.

The *Graf Zeppelin*

In August 1929 *Graf Zeppelin* departed for another daring enterprise: a circumnavigation of the globe. The growing popularity of the "giant of the air" made it easy for Eckener to find sponsors. One of these was the American press tycoon William Randolph Hearst, who requested that the tour officially start in Lakehurst. As with the October 1928 flight to New York, Hearst had placed a reporter, Grace Marguerite Hay Drummond-Hay, on board: she therefore became the first woman to circumnavigate the globe by air. From there, *Graf Zeppelin* flew to Friedrichshafen, then Tokyo, Los Angeles, and back to Lakehurst, in 21 days 5 hours and 31 minutes. Including the initial and final trips between Friedrichshafen and Lakehurst and back, the dirigible had travelled 49,618 kilometres (30,831 mi).

US Air Mail 1930 picturing Graf Zeppelin

In the following year, *Graf Zeppelin* undertook trips around Europe, and following a successful tour to Recife, Brazil in May 1930, it was decided to open the first regular transatlantic airship line. This line operated between Frankfurt and Recife, and was later extended to Rio de Janeiro, with a stop in Recife. Despite the beginning of the Great Depression and growing competition from fixed-wing aircraft, *LZ 127* transported an increasing volume of passengers and mail across the ocean every year until 1936. The ship made another spectacular voyage in July 1931 when it made a seven-day research trip to the Arctic.[N 2] This had already been a dream of Count von Zeppelin twenty years earlier, which could not be realized at the time due to the outbreak of war.

Eckener intended to follow the successful airship with another larger Zeppelin, designated LZ 128. This was to be powered by eight engines, 232 m (761 ft) in length, with a capacity of 199,980 m³ (7,062,100 cu ft). However the loss of the British passenger airship R101 on 5 October 1930 led the Zeppelin company to reconsider the safety of hydrogen-filled vessels, and the design was abandoned in favour of a new project, LZ 129. This was intended to be filled with inert helium.

Hindenburg, End of an Era

The coming to power of the Nazi Party in 1933 had important consequences for Zeppelin Luftschiffbau. Zeppelins became a propaganda tool for the new regime: they would now display the Nazi swastika on their fins and occasionally tour Germany to play march music and propaganda speeches to the people. In 1934 Joseph Goebbels, the Minister of Propaganda, contributed two million reichsmarks towards the construction of LZ 129 and in 1935 Hermann Göring established a new airline directed by Ernst Lehmann, the *Deutsche Zeppelin Reederei*, as a subsidiary of *Lufthansa* to take over Zeppelin operations. Hugo Eckener was an outspoken anti-Nazi: complaints about the use of Zeppelins for propaganda purposes in 1936 led Goebbels to declare "Dr. Eckener has placed himself outside the pale of society. Henceforth his name is not to be mentioned in the newspapers and his photograph is not to be published"

The *Hindenburg*: note swastikas on tail fins.

On 4 March 1936 LZ 129 *Hindenburg* (named after former President of Germany Paul von Hindenburg) made its first flight. The *Hindenburg* was the largest airship ever built. It had been designed to use non-flammable helium, but the only supplies of the gas were controlled by the United States, who refused to allow its export. So, in what proved to be a fatal decision, the *Hindenburg* was filled with flammable hydrogen. Apart from the propaganda missions, *LZ 129* was used on the transatlantic service alongside *Graf Zeppelin*.

The Hindenburg on fire in 1937

On 6 May 1937, while landing in Lakehurst after a transatlantic flight, the tail of the ship caught fire, and within seconds, the *Hindenburg* burst into flames, killing 35 of the 97 people on board and one member of the ground crew. The cause of the fire has not been definitively determined. The investigation into the accident concluded that static electricity had ignited hydrogen which had leaked from the gasbags, although there were allegations of sabotage. 13 passengers and 22 crew, including Ernst Lehmann, were killed.

Despite the apparent danger, there remained a list of 400 people who still wanted to fly as Zeppelin passengers and had paid for the trip. Their money was refunded in 1940.

Graf Zeppelin was retired one month after the *Hindenburg* wreck and turned into a museum. The intended new flagship Zeppelin was completed in 1938 and, inflated with hydrogen, made some test flights (the first on 14 September), but never carried passengers. Another project, *LZ 131*, designed to be even larger than *Hindenburg* and *Graf Zeppelin II*, never progressed beyond the production of a few ring frames.

Graf Zeppelin II was assigned to the *Luftwaffe* and made about 30 test flights prior to the beginning of World War II. Most of those flights were carried out near the Polish border, first in the Sudeten mountains region of Silesia, then in the Baltic Sea region. During one such flight *LZ 130* crossed the Polish border near the Hel Peninsula, where it was intercepted by a Polish Lublin R-XIII aircraft from Puck naval airbase and forced to leave Polish airspace. During this time, *LZ 130* was used for electronic scouting missions, and was equipped with various measuring equipment. In August 1939, it made a flight near the coastline of Great Britain in an attempt to determine whether the 100-metre towers erected from Portsmouth to Scapa Flow were used for aircraft radio location. Photography, radio wave interception, magnetic and radio frequency analysis were unable to detect operational British Chain Home radar due to searching in the wrong frequency range. The frequencies searched were too high, an assumption based on the Germans' own radar systems. The mistaken conclusion was the British towers were not connected with radar operations, but were for naval radio communications.

After the beginning of the Second World War on 1 September, the *Luftwaffe* ordered LZ 127 and *LZ 130* moved to a large Zeppelin hangar in Frankfurt, where the skeleton of LZ 131 was also located. In March 1940 Göring ordered the scrapping of the remaining airships, and on 6 May the Frankfurt hangars were demolished.

Cultural Influences

Zeppelins have been an inspiration to music, cinematography and literature. In 1934, the calypsonian Attila the Hun recorded "Graf Zeppelin", commemorating the airship's visit to Trinidad.

Zeppelins are often featured in alternate history fiction. In the American science fiction series, *Fringe*, Zeppelins are a notable historical idiosyncrasy that helps differentiate the series' two parallel universes, also used in *Doctor Who* in the episodes "The Rise of the Cybermen" and "The Age of Steel" when the TARDIS crashes in an alternate reality where Britain is a 'people's republic' and Pete Tyler, Rose Tyler's father, is alive and is a wealthy inventor. They are also seen in the alternate reality 1939 plot line in the film *Sky Captain and the World of Tomorrow*, and have an iconic association with the steampunk subcultural movement in broader terms. In 1989, Japanese animator Miyazaki released *Kiki's Delivery Service*, which features a Zeppelin as a plot element.

In 1968, English rock band Led Zeppelin chose their name after Keith Moon, drummer of The Who, told guitarist Jimmy Page that his idea to create a band would "go down like a lead balloon." Page's manager Peter Grant suggested changing the spelling of "Lead" to "Led" to avoid mispronunciation. "Balloon" was replaced with "Zeppelin" as Jimmy Page saw it as a symbol of "the perfect combination of heavy and light, combustibility and grace." For the group's self-titled debut album, Page suggested the group use a picture of the Hindenburg crashing in New Jersey in 1937, much to Frau Eva Von Zeppelin's disgust. Von Zeppelin tried to sue the group for using the name Zeppelin, but the case was eventually dismissed.

Modern era

Since the 1990s Zeppelin Luftschifftechnik, a daughter enterprise of the Zeppelin conglomerate that built the original German Zeppelins, has been developing Zeppelin "New Technology" (NT) airships. These vessels are semi-rigids based partly on internal pressure, partly on a frame.

The Airship Ventures company operated zeppelin passenger travel to California from October 2008 to November 2012 with one of these Zeppelin NT airships.

In May 2011, Goodyear announced that they will be replacing their fleet of blimps with Zeppelin NTs, resurrecting their partnership that ended over 70 years ago. They will also be building the airships in the United States. Modern zeppelins are held aloft by the inert gas helium, eliminating the danger of combustion illustrated by the *Hindenburg*. It has been proposed that modern zeppelins could be powered by hydrogen fuel cells. Often Zeppelin NTs are used for sightseeing trips, for example D-LZZF (c/n 03) was used for Edelweiss's birthday celebration performing flights over Austria, she is now used, weather permitting, on flights over Munich.

Monoplane

Supermarine Spitfire monoplane

A monoplane is a fixed-wing aircraft with a single main wing plane, in contrast to a biplane or other multiplane, each of which has multiple planes.

A monoplane has inherently the highest efficiency and lowest drag of any wing configuration and is the simplest to build. However, during the early years of flight these advantages were offset by its greater weight and lower manoeuvrability, making it relatively rare until the 1930s. Since then the monoplane has been the most common form for a fixed-wing aircraft.

History

The Santos-Dumont Demoiselle was the first production monoplane (replica shown).

Although the first successful aircraft were biplanes, the first attempts at heavier-than-air flying machines were monoplanes, and many pioneers continued to develop monoplane designs. For example, the first aeroplane to be put into production was the 1907 Santos-Dumont Demoiselle, while the Blériot XI flew across the English Channel in 1909. Throughout 1909–1910 Hubert Latham set multiple altitude records in his Antoinette IV monoplane, eventually reaching 1,384 m (4,541 ft).

The Junkers J1 monoplane pioneered all-metal construction in 1915.

The equivalent German language term is *Eindecker*, as in the mid-wing Fokker Eindecker fighter of 1915 which for a time dominated the skies in what became known as the "Fokker scourge". The German military Idflieg aircraft designation system prior to 1918 prefixed monoplane type designations with an *E*, until the approval of the Fokker D.VIII fighter from its former "E.V" designation. However the success of the Fokker was short-lived and World War One was dominated by biplanes. Towards the end of the war the parasol monoplane became popular and successful designs were produced into the 1920s.

Nonetheless, relatively few monoplane types were built between 1914 and the late 1920s, compared with the number of biplanes. The reasons for this were primarily practical. With the low engine powers and air speeds available, the wings of a monoplane needed to be large in order to create enough lift while a biplane could have two smaller wings and so be made smaller and lighter.

Towards the end of the first world war, the inherent high drag of the biplane was beginning to restrict performance. Engines were not yet powerful enough to make the heavy cantilever-wing monoplane viable and the braced parasol wing became popular on fighter aircraft, although few arrived in time to see combat. It remained popular throughout the 1920s.

On flying boats with a shallow hull, a parasol wing allows the engines to be mounted above the spray from the water when taking of and landing. It was popular on flying boats during the 1930s; a typical example being the Consolidated Catalina. It died out when taller hulls became the norm during WWII, allowing the wing to be attached directly to the hull.

As ever-increasing engine powers made the weight of all-metal construction and the cantilever wing more practical, they became common during the post-WWI period, the day of the braced wing passed, and by the 1930s the cantilever monoplane was fast becoming the standard configuration for a fixed-wing aircraft. Advanced monoplane fighter-aircraft designs were mass-produced for military services around the world in both the Soviet Union and the United States in the early-mid 1930s, with the Polikarpov I-16 and the Boeing P-26 Peashooter respectively.

Most military aircraft of WW2 were monoplanes, as have been virtually all aircraft since.

Jet and rocket engines have even more power and all modern high-speed aircraft, especially supersonic types, have been monoplanes.

Monoplane Characteristics

Support and Weight

The inherent efficiency of the monoplane can best be realised in the unbraced cantilever wing which carries all structural forces internally. By contrast a braced wing has additional drag from the exposed bracing struts and/or wires, leading to lower aerodynamic efficiency. On the other hand, the braced wing has greater structural efficiency and can be made much lighter. This in turn means that for a wing of a given size, bracing allows it to fly slower with a lower-powered engine, while a heavy cantilever wing needs a more powerful engine and can fly faster.

A cantilever wing can be made lighter by making it thicker. This increases internal storage for fuel, retractable undercarriage, armaments and in some cases even passengers and crew. A thick wing can also be given greater curvature of its upper surface and so create more lift than a thin wing, and some American bombers of WWII had unusually thick wings.

But thickness, like bracing, also increases drag, especially above the speed of sound. Supersonic aircraft have thin wings which are much heavier, and the wheels and fuel must find storage elsewhere.

Monoplane vs. Biplane

The earliest powered aircraft were biplanes, and this configuration was dominant until the mid-1920s, when the monoplane began to take over. Today, the biplane configuration is seldom used, although special types including aerobatic aircraft and kit planes are still available.

It might seem that a monoplane's wing should be double the area of the equivalent biplane wing, because the biplane has twice as many. This ought to make the monoplane wing larger and more unwieldy, but in practice the biplane's wings interfere with each other, making them less aerodynamically efficient and reducing any theoretical advantage.

A pair of biplane wings is typically braced with struts and wires to stiffen the structure and make it lighter, thereby enabling slow flight. However, even a strutted monoplane will still be aerodynamically more efficient than a biplane.

Human-powered aircraft, which are among the slowest and lightest of flying machines, are monoplanes with very large wings.

Wing Position

Besides the general variations in wing configuration such as tail position and use of bracing, the main distinction between types of monoplane is how high up the wings are mounted in relation to the fuselage.

Low Wing

Low wing on a Curtiss P-40

A low wing is one which is located on or near the base of the fuselage.

Placing the wing low down allows good visibility upwards and frees up the central fuselage from the wing spar carry-through. By reducing pendulum stability it makes the aircraft more manoeuvrable, as on the Spitfire; but aircraft that value stability over manoeuvrability may then need some dihedral. A low wing allows a lighter structure because the fuselage sides carry no additional loads, and the main undercarriage legs can be made shorter.

A feature of the low wing position is its significant ground effect, giving the plane a tendency to float further before landing. Conversely, this very ground effect permits shorter takeoffs.

The low wing configuration has proved particularly suitable for passenger jetliners.

Mid Wing

Mid wing on a de Havilland Vampire T11.

A mid wing is mounted midway up the fuselage. It is aerodynamically the cleanest and most balanced, but the carry-through spar structure can reduce the useful fuselage volume near its centre of gravity, where space is often in most demand. It is common on high-performance types such as sailplanes.

Shoulder Wing

Shoulder wing on an ARV Super2.

A shoulder wing (a category between high-wing and mid-wing) is a configuration whereby the wing is mounted near the top of the fuselage, but not on the very top. It is so called because it sits on the "shoulder" of the fuselage, rather than on the pilot's shoulder. Shoulder-wings and high-wings share some characteristics, namely: they support a pendulous fuselage which requires no wing dihedral for stability; and, by comparison with a low-wing, a shoulder-wing's limited ground effect reduces float on landing. Compared to a low-wing, shoulder-wing and high-wing configurations give increased propeller clearance on multi-engined aircraft.

On a large aircraft, there is little practical difference between a shoulder wing and a high wing; but on light aircraft the configuration is significant because it offers superior visibility to the pilot. On a light aircraft, the shoulder-wing may need to be swept forward to maintain correct center of gravity. Examples of light aircraft with shoulder wings include the ARV Super2, the Bölkow Junior and the Saab Safari.

High Wing

High wing on a de Havilland Canada Dash 8.

A high wing has its upper surface on or above the top of the fuselage. It shares many advantages and disadvantages with the shoulder wing, but on a light aircraft the high wing has poorer upwards visibility. On light aircraft such as the Cessna 152, the wing is usually located on top of the pilot's cabin, so that the centre of lift broadly coincides with the centre of gravity.

Parasol Wing

Parasol wing on a Pietenpol Air Camper.

A parasol wing aircraft is essentially a biplane without the lower pair of wings. The parasol wing is not directly attached to the fuselage, but is held above it, supported either by cabane struts or by a single pylon. Additional bracing may be provided by struts extending from the fuselage sides. Some early gliders had an open cockpit and a parasol wing mounted on a pylon.

The parasol wing was popular only during the inter-war transition years between biplanes and monoplanes. Compared to a biplane, a parasol wing has less bracing and lower drag; but compared to a high wing, there is so much extra drag that the parasol wing has become obsolete, although the Pietenpol Air Camper kitplane still has some aficionados.

Rotorcraft

A rotorcraft or rotary-wing aircraft is a heavier-than-air flying machine that uses lift generated by wings, called rotary wings or rotor blades, that revolve around a mast. Several rotor blades mounted on a single mast are referred to as a rotor. The International Civil Aviation Organization (ICAO) defines a rotorcraft as "supported in flight by the reactions of the air on one or more rotors". Rotorcraft generally include those aircraft where one or more rotors are required to provide lift throughout the entire flight, such as helicopters, cyclocopters, autogyros, and gyrodynes. Compound rotorcraft may also include additional thrust engines or propellers and static lifting surfaces.

An AS332 helicopter from the Hong Kong Government Flying Service conducts a water bomb demonstration

Classes of Rotorcraft

Helicopter

A helicopter is a rotorcraft whose rotors are driven by the engine(s) throughout the flight to allow the helicopter to take off vertically, hover, fly forwards, backwards and laterally, as well as to land vertically. Helicopters have several different configurations of one or more main rotors.

Helicopters with a single shaft-driven main lift rotor require some sort of antitorque device such as a tail rotor, fantail, or NOTAR, except some rare examples of helicopters using tip jet propulsion, which generates almost no torque.

Cyclogyro/Cyclocopter

A cyclocopter is a rotorcraft whose rotors are also driven by the engine throughout the flight, but the blades rotate about the horizontal axis while being parallel to it. They are being developed in a number of countries in order to replace the helicopters, which have a number of very serious shortcomings such as low efficiency in forward flight, low speed, very high noise and vibration levels, limited flight range, and low altitude ceiling. At the present time flying model prototypes have been built in China, US, S. Korea and Austria.

Autogyro

A German-registered autogyro

An autogyro (sometimes called gyrocopter, gyroplane, or rotaplane) utilizes an unpowered rotor, driven by aerodynamic forces in a state of autorotation to develop lift, and an engine-powered propeller, similar to that of a fixed-wing aircraft, to provide thrust. While similar to a helicopter rotor in appearance, the autogyro's rotor must have air flowing up and through the rotor disk in order to generate rotation. Early autogyros resembled the fixed-wing aircraft of the day, with wings and a front-mounted engine and propeller in a tractor configuration to pull the aircraft through the air. Late-model autogyros feature a rear-mounted engine and propeller in a pusher configuration.

The autogyro was invented in 1920 by Juan de la Cierva. The autogyro with pusher propeller was first tested by Etienne Dormoy with his Buhl A-1 Autogyro.

Gyrodyne

The rotor of a gyrodyne is normally driven by its engine for takeoff and landing—hovering like a helicopter—with anti-torque and propulsion for forward flight provided by one or more propellers mounted on short or stub wings. As power is increased to the propeller, less power is required by the rotor to provide forward thrust resulting in reduced pitch angles and rotor blade flapping. At cruise speeds with most or all of the thrust being provided by the propellers, the rotor receives power only sufficient to overcome the profile drag and maintain lift. The effect is a rotorcraft operating in a more efficient manner than the freewheeling rotor of an autogyro in autorotation, minimizing the adverse effects of retreating blade stall of helicopters at higher airspeeds.

Fairey Rotodyne prototype.

Rotor Kite

A rotor kite or gyroglider is an unpowered rotary-wing aircraft. Like an autogyro or helicopter, it relies on lift created by one or more sets of rotors in order to fly. Unlike a helicopter, autogyros and rotor kites do not have an engine powering their rotors, but while an autogyro has an engine providing forward thrust that keeps the rotor turning, a rotor kite has no engine at all, and relies on either being carried aloft and dropped from another aircraft, or by being towed into the air behind a car or boat.

Rotor Configuration

Number of Blades

A rotary wing is characterised by the number of blades. Typically this is between two and six per driveshaft.

Number of Rotors

A rotorcraft may have one or more rotors. Various rotor configurations have been used:

- One rotor. Powered rotors require compensation for the torque reaction causing yaw, except in the case of tipjet drive.

- Two rotors. These typically rotate in opposite directions cancelling the torque reaction so that no tail rotor or other yaw stabiliser is needed. These rotors can be laid out as

    o Tandem - One in front of the other.

    o Transverse - Side by side.

- o Coaxial - One rotor disc above the other, with concentric drive shafts.

- o Intermeshing rotors - Twin rotors at an acute angle from each other, whose nearly-vertical driveshafts are geared together to synchronise their rotor blades so that they intermesh, also called a **synchropter**.

- Three rotors. An uncommon configuration; the 1948 Cierva Air Horse had three rotors as it was not believed a single rotor of sufficient strength could be built for its size. All three rotors turned in the same direction and yaw compensation was provided by inclining each rotor axis to generate rotor thrust components that opposed torque.

- Four rotors. Also referred to as quadcopters/quadrotors, they typically have two rotors turning clockwise and two counter-clockwise.

- More than four rotors. These designs (referred to generally as multirotors, or sometimes individually as hexacopters and octocopters), have matched sets of rotors turning in opposite directions, and uncommon in full-size manned air-craft, but commonly seen in unmanned aerial vehicle systems.

Helicopter

A helicopter is a type of rotorcraft in which lift and thrust are supplied by rotors. This allows the helicopter to take off and land vertically, to hover, and to fly forward, back-ward, and laterally. These attributes allow helicopters to be used in congested or iso-lated areas where fixed-wing aircraft and many forms of VTOL (vertical takeoff and landing) aircraft cannot perform.

A US police Bell 206 helicopter

The English word *helicopter* is adapted from the French word *hélicoptère*, coined by Gustave Ponton d'Amécourt in 1861, which originates from the Greek *helix* (ἕλιξ) "helix, spiral, whirl, convolution" and *pteron* (πτερόν) "wing". English language nicknames for helicopter include "chopper", "copter", "helo", "heli", and "whirlybird".

Helicopters were developed and built during the first half-century of flight, with the Focke-Wulf Fw 61 being the first operational helicopter in 1936. Some helicopters reached limited production, but it was not until 1942 that a helicopter designed by Igor Sikorsky reached full-scale production, with 131 aircraft built. Though most earlier designs used more than one main rotor, it is the single main rotor with anti-torque tail rotor configuration that has become the most common helicopter configuration. Tandem rotor helicopters are also in widespread use due to their greater payload capacity. Coaxial helicopters, tiltrotor aircraft, and compound helicopters are all flying today. Quadcopter helicopters pioneered as early as 1907 in France, and other types of multicopter have been developed for specialized applications such as unmanned drones.

History

Early Design

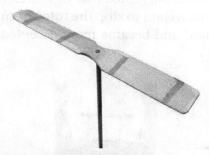

A decorated Japanese *taketombo* bamboo-copter

The earliest references for vertical flight came from China. Since around 400 BC, Chinese children have played with bamboo flying toys (or Chinese top). This bamboo-copter is spun by rolling a stick attached to a rotor. The spinning creates lift, and the toy flies when released. The 4th-century AD Daoist book *Baopuzi* by Ge Hong (抱朴子 "Master who Embraces Simplicity") reportedly describes some of the ideas inherent to rotary wing aircraft.

Designs similar to the Chinese helicopter toy appeared in Renaissance paintings and other works. In the 18th and early 19th centuries Western scientists developed flying machines based on the Chinese toy.

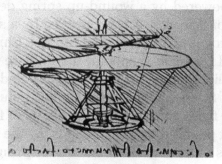

Leonardo's "aerial screw"

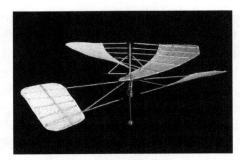

Experimental helicopter by Enrico Forlanini (1877), exposed at the Museo nazionale della scienza e della tecnologia Leonardo da Vinci of Milan

It was not until the early 1480s, when Leonardo da Vinci created a design for a machine that could be described as an "aerial screw", that any recorded advancement was made towards vertical flight. His notes suggested that he built small flying models, but there were no indications for any provision to stop the rotor from making the craft rotate. As scientific knowledge increased and became more accepted, men continued to pursue the idea of vertical flight.

Prototype created by M. Lomonosov, 1754

In July 1754, Russian Mikhail Lomonosov had developed a small coaxial modeled after the Chinese top but powered by a wound-up spring device and demonstrated it to the Russian Academy of Sciences. It was powered by a spring, and was suggested as a method to lift meteorological instruments. In 1783, Christian de Launoy, and his mechanic, Bienvenu, used a coaxial version of the Chinese top in a model consisting of contrarotating turkey flight feathers as rotor blades, and in 1784, demonstrated it to the French Academy of Sciences. Sir George Cayley, influenced by a childhood fascination with the Chinese flying top, developed a model of feathers, similar to that of Launoy and Bienvenu, but powered by rubber bands. By the end of the century, he had progressed to using sheets of tin for rotor blades and springs for power. His writings on his experiments and models would become influential on future aviation pioneers.

Alphonse Pénaud would later develop coaxial rotor model helicopter toys in 1870, also powered by rubber bands. One of these toys, given as a gift by their father, would inspire the Wright brothers to pursue the dream of flight.

In 1861, the word "helicopter" was coined by Gustave de Ponton d'Amécourt, a French inventor who demonstrated a small steam-powered model. While celebrated as an innovative use of a new metal, aluminum, the model never lifted off the ground. D'Amecourt's linguistic contribution would survive to eventually describe the vertical flight he had envisioned. Steam power was popular with other inventors as well. In 1878 the Italian Enrico Forlanini's unmanned vehicle, also powered by a steam engine, rose to a height of 12 meters (40 ft), where it hovered for some 20 seconds after a vertical take-off. Emmanuel Dieuaide's steam-powered design featured counter-rotating rotors powered through a hose from a boiler on the ground. In 1887 Parisian inventor, Gustave Trouvé, built and flew a tethered electric model helicopter.

On July 1901, Hermann Ganswindt demonstrated maiden flight of his helicopter took place in Berlin-Schöneberg, which probably was the first motor-driven flight carrying humans. A movie covering the event was taken by Max Skladanowsky, but it remains lost.

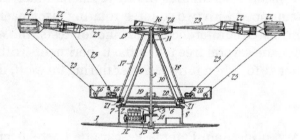

Drawing from Edison's 1910 patent

In 1885, Thomas Edison was given US$1,000 by James Gordon Bennett, Jr., to conduct experiments towards developing flight. Edison built a helicopter and used the paper for a stock ticker to create guncotton, with which he attempted to power an internal combustion engine. The helicopter was damaged by explosions and one of his workers was badly burned. Edison reported that it would take a motor with a ratio of three to four pounds per horsepower produced to be successful, based on his experiments. Ján Bahýľ, a Slovak inventor, adapted the internal combustion engine to power his helicopter model that reached a height of 0.5 meters (1.6 ft) in 1901. On 5 May 1905, his helicopter reached four meters (13 ft) in altitude and flew for over 1,500 meters (4,900 ft). In 1908, Edison patented his own design for a helicopter powered by a gasoline engine with box kites attached to a mast by cables for a rotor, but it never flew.

First Flights

In 1906, two French brothers, Jacques and Louis Breguet, began experimenting with airfoils for helicopters. In 1907, those experiments resulted in the *Gyroplane No.1*, possibly as the earliest known example of a quadcopter. Although there is some uncertainty about the date, sometime between 14 August and 29 September 1907, the Gyroplane No. 1 lifted its pilot into the air about two feet (0.6 m) for a minute. The Gyroplane No. 1 proved to be extremely unsteady and required a man at each corner of the airframe to hold it steady. For this reason, the flights of the Gyroplane No. 1 are considered to be the first manned flight of a helicopter, but not a free or untethered flight.

That same year, fellow French inventor Paul Cornu designed and built a Cornu helicopter that used two 20-foot (6 m) counter-rotating rotors driven by a 24 hp (18 kW) Antoinette engine. On 13 November 1907, it lifted its inventor to 1 foot (0.3 m) and remained aloft for 20 seconds. Even though this flight did not surpass the flight of the Gyroplane No. 1, it was reported to be the first truly free flight with a pilot.[n 1] Cornu's helicopter completed a few more flights and achieved a height of nearly 6.5 feet (2 m), but it proved to be unstable and was abandoned.

In 1911, Slovenian philosopher and economist Ivan Slokar patented a helicopter configuration.

The Danish inventor Jacob Ellehammer built the Ellehammer helicopter in 1912. It consisted of a frame equipped with two counter-rotating discs, each of which was fitted with six vanes around its circumference. After indoor tests, the aircraft was demonstrated outdoors and made several free take-offs. Experiments with the helicopter continued until September 1916, when it tipped over during take-off, destroying its rotors.

Early Development

In the early 1920s, Argentine Raúl Pateras-Pescara de Castelluccio, while working in Europe, demonstrated one of the first successful applications of cyclic pitch. Coaxial, contra-rotating, biplane rotors could be warped to cyclically increase and decrease the lift they produced. The rotor hub could also be tilted forward a few degrees, allowing the aircraft to move forward without a separate propeller to push or pull it. Pateras-Pescara was also able to demonstrate the principle of autorotation. By January 1924, Pescara's helicopter No. 1 was tested but was found to be underpowered and could not lift its own weight. His 2F fared better and set a record. The British government funded further research by Pescara which resulted in helicopter No. 3, powered by a 250 hp radial engine which could fly for up to ten minutes.

Oehmichen N°2, 1923

On 14 April 1924 Frenchman Étienne Oehmichen set the first helicopter world record recognized by the *Fédération Aéronautique Internationale* (FAI), flying his quadrotor helicopter 360 meters (1,181 ft). On 18 April 1924, Pescara beat Oemichen's record, flying for a distance of 736 meters (nearly a half mile) in 4 minutes and 11 seconds (about 8 mph, 13 km/h), maintaining a height of six feet (1.8 meters). On 4 May, Oehmichen set the first 1 km closed-circuit helicopter flight in 7 minutes 40 seconds with his No. 2 machine.

In the US, George de Bothezat built the quadrotor helicopter de Bothezat helicopter for the United States Army Air Service but the Army cancelled the program in 1924, and the aircraft was scrapped.

Albert Gillis von Baumhauer, a Dutch aeronautical engineer, began studying rotorcraft design in 1923. His first prototype "flew" ("hopped" and hovered in reality) on 24 September 1925, with Dutch Army-Air arm Captain Floris Albert van Heijst at the controls. The controls that van Heijst used were von Baumhauer's inventions, the cyclic and collective. Patents were granted to von Baumhauer for his cyclic and collective controls by the British ministry of aviation on 31 January 1927, under patent number 265,272.

In 1928, Hungarian aviation engineer Oszkár Asbóth constructed a helicopter prototype that took off and landed at least 182 times, with a maximum single flight duration of 53 minutes.

In 1930, the Italian engineer Corradino D'Ascanio built his D'AT3, a coaxial helicopter. His relatively large machine had two, two-bladed, counter-rotating rotors. Control was achieved by using auxiliary wings or servo-tabs on the trailing edges of the blades, a concept that was later adopted by other helicopter designers, including Bleeker and Kaman. Three small propellers mounted to the airframe were used for additional pitch, roll, and yaw control. The D'AT3 held modest FAI speed and altitude records for the time, including altitude (18 m or 59 ft), duration (8 minutes 45 seconds) and distance flown (1,078 m or 3,540 ft).

In the Soviet Union, Boris N. Yuriev and Alexei M. Cheremukhin, two aeronautical engineers working at the *Tsentralniy Aerogidrodinamicheskiy Institut* (TsAGI, the Central Aerohydrodynamic Institute), constructed and flew the TsAGI 1-EA single lift-rotor helicopter, which used an open tubing framework, a four-blade main lift rotor, and twin sets of 1.8-meter (6-foot) diameter, two-bladed anti-torque rotors: one set of two at the nose and one set of two at the tail. Powered by two M-2 powerplants, up-rated copies of the Gnome *Monosoupape* 9 Type B-2 100 CV output rotary engine of World War I, the TsAGI 1-EA made several low altitude flights. By 14 August 1932, Cheremukhin managed to get the 1-EA up to an unofficial altitude of 605 meters (1,985 ft), shattering d'Ascanio's earlier achievement. As the Soviet Union was not yet a member of the FAI, however, Cheremukhin's record remained unrecognized.

Nicolas Florine, a Russian engineer, built the first twin tandem rotor machine to perform a free flight. It flew in Sint-Genesius-Rode, at the *Laboratoire Aérotechnique de Belgique* (now von Karman Institute) in April 1933, and attained an altitude of six meters (20 ft) and an endurance of eight minutes. Florine chose a co-rotating configuration because the gyroscopic stability of the rotors would not cancel. Therefore, the rotors had to be tilted slightly in opposite directions to counter torque. Using hingeless rotors and co-rotation also minimised the stress on the hull. At the time, it was one of the most stable helicopters in existence.

The Bréguet-Dorand *Gyroplane Laboratoire* was built in 1933. It was a coaxial helicopter, contra-rotating. After many ground tests and an accident, it first took flight on 26 June 1935. Within a short time, the aircraft was setting records with pilot Maurice Claisse at the controls. On 14 December 1935, he set a record for closed-circuit flight with a 500-meter (1,600 ft) diameter. The next year, on 26 September 1936, Claisse set a height record of 158 meters (520 ft). And, finally, on 24 November 1936, he set a flight duration record of one hour, two minutes and 50 seconds over a 44 kilometer (27 mi) closed circuit at 44.7 kilometers per hour (27.8 mph). The aircraft was destroyed in 1943 by an Allied airstrike at Villacoublay airport.

Arthur M. Young, American inventor, started work on model helicopters in 1928 using converted electric hover motors to drive the rotor head. Young invented the stabilizer bar and patented it shortly after. A mutual friend introduced Young to Lawrence Dale, who once seeing his work asked him to join the Bell Aircraft company. When Young arrived at Bell in 1941, he signed his patent over and began work on the helicopter. His budget was US$250,000 to build 2 working helicopters. In just 6 months they completed the first Bell Model 1, which spawned the Bell Model 30, later succeeded by the Bell 47.

Autogyro

Pitcairn PCA-2 autogyro, built in the U.S. under licence to the Cierva Autogiro Company

Early rotor winged flight suffered failures primarily associated with the unbalanced rolling movement generated when attempting take-off, due to dissymmetry of lift between the advancing and retreating blades. This major difficulty was resolved by Juan de la Cierva's introduction of the flapping hinge. In 1923, de la Cierva's first successful autogyro was flown in Spain by Lt. Gomez Spencer. In 1925 he brought his C.6 to Britain and demonstrated it to the Air Ministry at Farnborough, Hampshire. This machine had a four blade rotor with flapping hinges but relied upon conventional airplane controls for pitch, roll and yaw. It was based upon an Avro 504K fuselage, initial rotation of the rotor was achieved by the rapid uncoiling of a rope passed around stops on the undersides of the blades.

A major problem with the autogyro was driving the rotor before takeoff. Several methods were attempted in addition to the coiled rope system, which could take the rotor speed to 50% of that required, at which point movement along the ground to reach flying speed was necessary, while tilting the rotor to establish autorotation. Another approach was to tilt the tail stabiliser to deflect engine slipstream up through the rotor. The most acceptable solution was finally achieved with the C.19 Mk.4, which was produced in some quantities; a direct drive from the engine to the rotor was fitted, through which the rotor could be accelerated up to speed. The system was then declutched before the take-off run.

As de la Cierva's autogyros achieved success and acceptance, others began to follow and with them came further innovation. Most important was the development of direct rotor control through cyclic pitch variation, achieved initially by tilting the rotor hub and subsequently by the Austrian engineer Raoul Hafner, by the application of a spider mechanism that acted directly on each rotor blade. The first production direct control autogyro was the C.30, produced in quantity by Avro, Liore et Olivier, and Focke-Wulf.

The production model, called the C.30A by Avro, was built under licence in Britain, France and Germany and was similar to the C.30P. It carried small movable trimming surfaces. Each licensee used nationally built engines and used slightly different names. In all, 143 production C.30s were built, making it by far the most numerous pre-war autogyro.

Between 1933 and 1936, de la Cierva used one C.30A (*G-ACWF*) to perfect his last contribution to autogyro development before his death in late 1936. To enable the aircraft to take off without forward ground travel, he produced the "Autodynamic" rotor head, which allowed the rotor to be spun up by the engine in the usual way but to higher than take-off r.p.m at zero rotor incidence and then to reach operational positive pitch suddenly enough to jump some 20 ft (6 m) upwards.

Birth of an Industry

Heinrich Focke at Focke-Wulf was licensed to produce the Cierva C.30 autogyro in 1933. Focke designed the world's first practical transverse twin-rotor helicopter, the Focke-Wulf Fw 61, which first flew on 26 June 1936. The Fw 61 broke all of the helicopter world records in 1937, demonstrating a flight envelope that had only previously been achieved by the autogyro.

Igor Sikorsky and the world's first mass-produced helicopter, the Sikorsky R-4, 1944

First airmail service by helicopter in Los Angeles, 1947

During World War II, Nazi Germany used helicopters in small numbers for observation, transport, and medical evacuation. The Flettner Fl 282 *Kolibri* synchropter—using the same basic configuration as Anton Flettner's own pioneering Fl 265—was used in the Mediterranean, while the Focke Achgelis Fa 223 *Drache* twin-rotor helicopter was used in Europe. Extensive bombing by the Allied forces prevented Germany from producing any helicopters in large quantities during the war.

In the United States, Russian-born engineer Igor Sikorsky and W. Lawrence LePage competed to produce the U.S. military's first helicopter. LePage received the patent rights to develop helicopters patterned after the Fw 61, and built the XR-1. Meanwhile, Sikorsky settled on a simpler, single rotor design, the VS-300, which turned out to be the first practical single lifting-rotor helicopter design. After experimenting with configurations to counteract the torque produced by the single main rotor, Sikorsky settled on a single, smaller rotor mounted on the tailboom.

Developed from the VS-300, Sikorsky's R-4 was the first large-scale mass-produced helicopter, with a production order for 100 aircraft. The R-4 was the only Allied helicopter to serve in World War II, when it was used primarily for search and rescue (by the USAAF 1st Air Commando Group) in Burma; in Alaska; and in other areas with harsh terrain. Total production reached 131 helicopters before the R-4 was replaced by other Sikorsky helicopters such as the R-5 and the R-6. In all, Sikorsky produced over 400 helicopters before the end of World War II.

While LePage and Sikorsky built their helicopters for the military, Bell Aircraft hired Arthur Young to help build a helicopter using Young's two-blade teetering rotor design, which used a weighted stabilizer bar placed at a 90° angle to the rotor blades. The subsequent Model 30 helicopter showed the design's simplicity and ease of use. The Model 30 was developed into the Bell 47, which became the first helicopter certified for civilian use in the United States. Produced in several countries, the Bell 47 was the most popular helicopter model for nearly 30 years.

Turbine Age

In 1951, at the urging of his contacts at the Department of the Navy, Charles Kaman modified his K-225 synchropter — a design for a twin-rotor helicopter concept first pioneered by Anton Flettner in 1939, with the aforementioned Fl 265 piston-engined design in Germany — with a new kind of engine, the turboshaft engine. This adaptation of the turbine engine provided a large amount of power to Kaman's helicopter with a lower weight penalty than piston engines, with their heavy engine blocks and auxiliary components. On 11 December 1951, the Kaman K-225 became the first turbine-powered helicopter in the world. Two years later, on 26 March 1954, a modified Navy HTK-1, another Kaman helicopter, became the first twin-turbine helicopter to fly. However, it was the Sud Aviation Alouette II that would become the first helicopter to be produced with a turbine-engine.

Reliable helicopters capable of stable hover flight were developed decades after fixed-wing aircraft. This is largely due to higher engine power density requirements than fixed-wing aircraft. Improvements in fuels and engines during the first half of the 20th century were a critical factor in helicopter development. The availability of lightweight turboshaft engines in the second half of the 20th century led to the development of larger, faster, and higher-performance helicopters. While smaller and less expensive helicopters still use piston engines, turboshaft engines are the preferred powerplant for helicopters today.

Uses

Due to the operating characteristics of the helicopter—its ability to take off and land vertically, and to hover for extended periods of time, as well as the aircraft's handling properties under low airspeed conditions—it has been chosen to conduct tasks that were previously not possible with other aircraft, or were time- or work-intensive to accomplish on the ground. Today, helicopter uses include transportation of people and cargo, military uses, construction, firefighting, search and rescue, tourism, medical transport, law enforcement, agriculture, news and media, and aerial observation, among others.

A helicopter used to carry loads connected to long cables or slings is called an aerial crane. Aerial cranes are used to place heavy equipment, like radio transmission towers and large air conditioning units, on the tops of tall buildings, or when an item must be raised up in a remote area, such as a radio tower raised on the top of a hill or mountain. Helicopters are used as aerial cranes in the logging industry to lift trees out of terrain where vehicles cannot travel and where environmental concerns prohibit the building of roads. These operations are referred to as longline because of the long, single sling line used to carry the load.

The largest single non-combat helicopter operation in history was the disaster management operation following the 1986 Chernobyl nuclear disaster. Hundreds of pilots were involved in airdrop and observation missions, making dozens of sorties a day for several months.

"Helitack" is the use of helicopters to combat wildland fires. The helicopters are used for aerial firefighting (water bombing) and may be fitted with tanks or carry helibuckets. Helibuckets, such as the Bambi bucket, are usually filled by submerging the bucket into lakes, rivers, reservoirs, or portable tanks. Tanks fitted onto helicopters are filled from a hose while the helicopter is on the ground or water is siphoned from lakes or reservoirs through a hanging snorkel as the helicopter hovers over the water source. Helitack helicopters are also used to deliver firefighters, who rappel down to inaccessible areas, and to resupply firefighters. Common firefighting helicopters include variants of the Bell 205 and the Erickson S-64 Aircrane helitanker.

Helicopters are used as air ambulances for emergency medical assistance in situations when an ambulance cannot easily or quickly reach the scene, or cannot transport the patient to a medical facility in time. Helicopters are also used when patients need to be transported between medical facilities and air transportation is the most practical method. An air ambulance helicopter is equipped to stabilize and provide limited medical treatment to a patient while in flight. The use of helicopters as air ambulances is often referred to as "MEDEVAC", and patients are referred to as being "airlifted", or "medevaced". This use was pioneered in the Korean war, when time to reach a medical facility was reduced to three hours from the eight hours needed in World War II, and further reduced to two hours by the Vietnam war.

Police departments and other law enforcement agencies use helicopters to pursue suspects. Since helicopters can achieve a unique aerial view, they are often used in conjunction with police on the ground to report on suspects' locations and movements. They are often mounted with lighting and heat-sensing equipment for night pursuits.

Military forces use attack helicopters to conduct aerial attacks on ground targets. Such helicopters are mounted with missile launchers and miniguns. Transport helicopters are used to ferry troops and supplies where the lack of an airstrip would make transport via fixed-wing aircraft impossible. The use of transport helicopters to deliver troops as an attack force on an objective is referred to as "air assault". Unmanned aerial systems (UAS) helicopter systems of varying sizes are developed by companies for military reconnaissance and surveillance duties. Naval forces also use helicopters equipped with dipping sonar for anti-submarine warfare, since they can operate from small ships.

Oil companies charter helicopters to move workers and parts quickly to remote drilling sites located at sea or in remote locations. The speed advantage over boats makes the high operating cost of helicopters cost-effective in ensuring that oil platforms continue to operate. Various companies specialize in this type of operation.

Other uses of helicopters include:

- Aerial photography
- Motion picture photography
- Electronic news gathering
- Reflection seismology
- Search and rescue
- Tourism and recreation
- Transport

Design Features

Rotor System

A teetering rotor system with a weighted flybar device

The rotor system, or more simply *rotor*, is the rotating part of a helicopter that generates lift. A rotor system may be mounted horizontally, as main rotors are, providing lift vertically, or it may be mounted vertically, such as a tail rotor, to provide horizontal thrust to counteract torque from the main rotors. The rotor consists of a mast, hub and rotor blades.

The mast is a cylindrical metal shaft that extends upwards from the transmission. At the top of the mast is the attachment point for the rotor blades called the hub. The rotor blades are attached to the hub. Main rotor systems are classified according to how the rotor blades are attached and move relative to the hub. There are three basic types: hingeless, fully articulated, and teetering; although some modern rotor systems use a combination of these.

Anti-torque Features

MD Helicopters 520N NOTAR

Most helicopters have a single main rotor, but torque created as the engine turns the rotor causes the body of the helicopter to turn in the opposite direction to the rotor (by conservation of angular momentum). To eliminate this effect, some sort of anti-torque control must be used.

The design that Igor Sikorsky settled on for his VS-300 was a smaller tail rotor. The tail rotor pushes or pulls against the tail to counter the torque effect, and this has become the most common configuration for helicopter design.

Some helicopters use other anti-torque controls instead of the tail rotor, such as the ducted fan (called *Fenestron* or *FANTAIL*) and NOTAR. NOTAR provides anti-torque similar to the way a wing develops lift through the use of the Coandă effect on the tail-boom.

Boeing CH-47 Chinook is the most common dual rotor helicopter deployed today

The use of two or more horizontal rotors turning in opposite directions is another configuration used to counteract the effects of torque on the aircraft without relying on an anti-torque tail rotor. This allows the power normally required to drive the tail rotor to be applied to the main rotors, increasing the aircraft's lifting capacity. There are several common configurations that use the counter-rotating effect to benefit the rotorcraft:

- Tandem rotors are two counter-rotating rotors with one mounted behind the other.

- Coaxial rotors are two counter-rotating rotors mounted one above the other with the same axis.

- Intermeshing rotors are two counter-rotating rotors mounted close to each other at a sufficient angle to let the rotors intermesh over the top of the aircraft without colliding.

- Transverse rotors are pair of counter-rotating rotors mounted at each end of the wings or outrigger structures. They are found on tiltrotors and some earlier helicopters.

- Quadcopters have four rotors often with parallel axes (sometimes rotating in the same direction with tilted axes) which are commonly used on model aircraft.

Tip jet designs let the rotor push itself through the air and avoid generating torque.

Engines

A turbine engine used in the Mi-2 helicopter.

The number, size and type of engine(s) used on a helicopter determines the size, function and capability of that helicopter design. The earliest helicopter engines were simple mechanical devices, such as rubber bands or spindles, which relegated the size of helicopters to toys and small models. For a half century before the first airplane flight, steam engines were used to forward the development of the understanding of helicopter aerodynamics, but the limited power did not allow for manned flight. The introduction of the internal combustion engine at the end of the 19th century became the watershed for helicopter development as engines began to be developed and produced that were powerful enough to allow for helicopters able to lift humans.

Early helicopter designs utilized custom-built engines or rotary engines designed for airplanes, but these were soon replaced by more powerful automobile engines and radial engines. The single, most-limiting factor of helicopter development during the first half of the 20th century was that the amount of power produced by an engine was not able to overcome the engine's weight in vertical flight. This was overcome in early successful helicopters by using the smallest engines available. When the compact, flat engine was developed, the helicopter industry found a lighter-weight powerplant easily adapted to small helicopters, although radial engines continued to be used for larger helicopters.

Turbine engines revolutionized the aviation industry, and the turboshaft engine finally gave helicopters an engine with a large amount of power and a low weight penalty. Turboshafts are also more reliable than piston engines, especially when producing the sustained high levels of power required by a helicopter. The turboshaft engine was able to be scaled to the size of the helicopter being designed, so that all but the lightest of helicopter models are powered by turbine engines today.

Special jet engines developed to drive the rotor from the rotor tips are referred to as tip jets. Tip jets powered by a remote compressor are referred to as cold tip jets, while those powered by combustion exhaust are referred to as hot tip jets. An example of a cold jet helicopter is the Sud-Ouest Djinn, and an example of the hot tip jet helicopter is the YH-32 Hornet.

Some radio-controlled helicopters and smaller, helicopter-type unmanned aerial vehicles, use electric motors. Radio-controlled helicopters may also have piston engines that use fuels other than gasoline, such as nitromethane. Some turbine engines commonly used in helicopters can also use biodiesel instead of jet fuel.

There are also human-powered helicopters.

Flight Controls

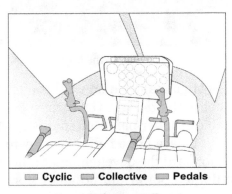

Controls from a Bell 206

A helicopter has four flight control inputs. These are the cyclic, the collective, the anti-torque pedals, and the throttle. The cyclic control is usually located between the pilot's legs and is commonly called the *cyclic stick* or just *cyclic*. On most helicopters, the cyclic is similar to a joystick. However, the Robinson R22 and Robinson R44 have a unique teetering bar cyclic control system and a few helicopters have a cyclic control that descends into the cockpit from overhead.

The control is called the cyclic because it changes the pitch of the rotor blades cyclically. The result is to tilt the rotor disk in a particular direction, resulting in the helicopter moving in that direction. If the pilot pushes the cyclic forward, the rotor disk tilts forward, and the rotor produces a thrust in the forward direction. If the pilot pushes the cyclic to the side, the rotor disk tilts to that side and produces thrust in that direction, causing the helicopter to hover sideways.

The collective pitch control or *collective* is located on the left side of the pilot's seat with a settable friction control to prevent inadvertent movement. The collective changes the pitch angle of all the main rotor blades collectively (i.e. all at the same time) and independently of their position. Therefore, if a collective input is made, all the blades change equally, and the result is the helicopter increasing or decreasing in altitude.

The anti-torque pedals are located in the same position as the rudder pedals in a fixed-wing aircraft, and serve a similar purpose, namely to control the direction in which the nose of the aircraft is pointed. Application of the pedal in a given direction changes the pitch of the tail rotor blades, increasing or reducing the thrust produced by the tail rotor and causing the nose to yaw in the direction of the applied pedal. The pedals mechanically change the pitch of the tail rotor altering the amount of thrust produced.

Helicopter rotors are designed to operate in a narrow range of RPM. The throttle controls the power produced by the engine, which is connected to the rotor by a fixed ratio transmission. The purpose of the throttle is to maintain enough engine power to keep the rotor RPM within allowable limits so that the rotor produces enough lift for flight. In single-engine helicopters, the throttle control is a motorcycle-style twist grip mounted on the collective control, while dual-engine helicopters have a power lever for each engine.

A swashplate controls the collective and cyclic pitch of the main blades. The swashplate moves up and down, along the main shaft, to change the pitch of both blades. This causes the helicopter to push air downward or upward, depending on the angle of attack. The swashplate can also change its angle to move the blades angle forwards or backwards, or left and right, to make the helicopter move in those directions.

Flight

There are three basic flight conditions for a helicopter: hover, forward flight and the transition between the two.

Hover

Hovering is the most challenging part of flying a helicopter. This is because a helicopter generates its own gusty air while in a hover, which acts against the fuselage and flight control surfaces. The end result is constant control inputs and corrections by the pilot to keep the helicopter where it is required to be. Despite the complexity of the task, the control inputs in a hover are simple. The cyclic is used to eliminate drift in the horizontal plane, that is to control forward and back, right and left. The collective is used to maintain altitude. The pedals are used to control nose direction or heading. It is the interaction of these controls that makes hovering so difficult, since an adjustment in any one control requires an adjustment of the other two, creating a cycle of constant correction.

Transition from Hover to forward Flight

As a helicopter moves from hover to forward flight it enters a state called translational lift which provides extra lift without increasing power. This state, most typically, occurs when the airspeed reaches approximately 16–24 knots, and may be necessary for a helicopter to obtain flight.

Forward Flight

In forward flight a helicopter's flight controls behave more like those of a fixed-wing aircraft. Displacing the cyclic forward will cause the nose to pitch down, with a resultant increase in airspeed and loss of altitude. Aft cyclic will cause the nose to pitch up, slowing the helicopter and causing it to climb. Increasing collective (power) while maintaining a constant airspeed will induce a climb while decreasing collective will cause a descent. Coordinating these two inputs, down collective plus aft cyclic or up collective plus forward cyclic, will result in airspeed changes while maintaining a constant altitude. The pedals serve the same function in both a helicopter and a fixed-wing aircraft, to maintain balanced flight. This is done by applying a pedal input in whichever direction is necessary to center the ball in the turn and bank indicator.

Safety

HAL Dhruv at the 2008 Royal International Air Tattoo, England

Royal Australian Navy Squirrel helicopters during a display at the 2008 Melbourne Grand Prix

A Robinson R44 Raven II arrives for the 2014 Royal International Air Tattoo, England

Maximum Speed Limit

The main limitation of the helicopter is its low speed. There are several reasons a helicopter cannot fly as fast as a fixed-wing aircraft. When the helicopter is hovering, the outer tips of the rotor travel at a speed determined by the length of the blade and the rotational speed. In a moving helicopter, however, the speed of the blades relative to the air depends on the speed of the helicopter as well as on their rotational speed. The airspeed of the advancing rotor blade is much higher than that of the helicopter itself. It is possible for this blade to exceed the speed of sound, and thus produce vastly increased drag and vibration.

At the same time, the advancing blade creates more lift traveling forward, the retreating blade produces less lift. If the aircraft were to accelerate to the air speed that the blade tips are spinning, the retreating blade passes through air moving at the same speed of the blade and produces no lift at all, resulting in very high torque stresses on the central shaft that can tip down the retreating-blade side of the vehicle, and cause a loss of control. Dual counter-rotating blades prevent this situation due to having two advancing and two retreating blades with balanced forces.

Because the advancing blade has higher airspeed than the retreating blade and generates a dissymmetry of lift, rotor blades are designed to "flap" – lift and twist in such a way that the advancing blade flaps up and develops a smaller angle of attack. Conversely, the retreating blade flaps down, develops a higher angle of attack, and generates more lift. At high speeds, the force on the rotors is such that they "flap" excessively, and the retreating blade can reach too high an angle and stall. For this reason, the maximum safe forward airspeed of a helicopter is given a design rating called V_{NE}, *velocity, never exceed*. In addition, it is possible for the helicopter to fly at an airspeed where an excessive amount of the retreating blade stalls, which results in high vibration, pitch-up, and roll into the retreating blade.

Noise

During the closing years of the 20th century designers began working on helicopter noise reduction. Urban communities have often expressed great dislike of noisy aircraft, and police and passenger helicopters can be unpopular. The redesigns followed the closure of some city heliports and government action to constrain flight paths in national parks and other places of natural beauty.

Vibration

Helicopters also vibrate; an unadjusted helicopter can easily vibrate so much that it will shake itself apart. To reduce vibration, all helicopters have rotor adjustments for height and weight. Blade height is adjusted by changing the pitch of the blade. Weight is adjusted by adding or removing weights on the rotor head and/or at the blade end caps. Most also have vibration dampers for height and pitch. Some also use mechanical feedback systems to sense and counter vibration. Usually the feedback system uses a mass as a "stable reference" and a linkage from the mass operates a flap to adjust the rotor's angle of attack to counter the vibration. Adjustment is difficult in part because measurement of the vibration is hard, usually requiring sophisticated accelerometers mounted throughout the airframe and gearboxes. The most common blade vibration adjustment measurement system is to use a stroboscopic flash lamp, and observe painted markings or coloured reflectors on the underside of the rotor blades. The traditional low-tech system is to mount coloured chalk on the rotor tips, and see how they mark a linen sheet. Gearbox vibration most often requires a gearbox overhaul or replacement. Gearbox or drive train vibrations can be extremely harmful to a pilot. The most severe being pain, numbness, loss of tactile discrimination and dexterity.

Loss of tail-rotor Effectiveness

For a standard helicopter with a single main rotor, the tips of the main rotor blades produce a vortex ring in the air, which is a spiraling and circularly rotating airflow. As the craft moves forward, these vortices trail off behind the craft.

When hovering with a forward diagonal crosswind, or moving in a forward diagonal direction, the spinning vortices trailing off the main rotor blades will align with the rotation of the tail rotor and cause an instability in flight control.

When the trailing vortices colliding with the tail rotor are rotating in the same direction, this causes a loss of thrust from the tail rotor. When the trailing vortices rotate in the opposite direction of the tail rotor, thrust is increased. Use of the foot pedals is required to adjust the tail rotor's angle of attack, to compensate for these instabilities.

These issues are due to the exposed tail rotor cutting through open air around rear of the vehicle. This issue disappears when the tail is instead ducted, using an internal impeller enclosed in the tail and a jet of high pressure air sideways out of the tail, as the main rotor vortices can not impact the operation of an internal impeller.

Critical Wind Azimuth

For a standard helicopter with a single main rotor, maintaining steady flight with a crosswind presents an additional flight control problem, where strong crosswinds from certain angles will increase or decrease lift from the main rotors. This effect is also triggered in a no-wind condition when moving the craft diagonally in various directions, depending on the direction of main rotor rotation.

This can lead to a loss of control and a crash or hard landing when operating at low altitudes, due to the sudden unexpected loss of lift, and insufficient time and distance available to recover.

Transmission

Pascal Chretien hovering the world's first manned electric helicopter, August 2011

Conventional rotary-wing aircraft use a set of complex mechanical gearboxes to convert the high rotation speed of gas turbines into the low speed required to drive main and tail rotors. Unlike powerplants, mechanical gearboxes cannot be duplicated (for redundancy) and have always been a major weak point in helicopter reliability. In-flight catastrophic gear failures often result in gearbox jamming and subsequent fatalities, whereas loss of lubrication can trigger onboard fire. Another weakness of mechanical gearboxes is their transient power limitation, due to structural fatigue limits. Recent EASA studies point to engines and transmissions as prime cause of crashes just after pilot errors.

By contrast, electromagnetic transmissions do not use any parts in contact; hence lubrication can be drastically simplified, or eliminated. Their inherent redundancy offers good resilience to single point of failure. The absence of gears enables high power transient without impact on service life. The concept of electric propulsion applied to helicopter and electromagnetic drive was brought to reality by Pascal Chretien who designed, built and flew world's first man-carrying, free-flying electric helicopter. The concept was taken from the conceptual computer-aided design model on September 10, 2010 to the first testing at 30% power on March 1, 2011 - less than six months. The aircraft first flew on August 12, 2011. All development was conducted in Venelles, France.

Hazards

As with any moving vehicle, unsafe operation could result in loss of control, structural damage, or loss of life. The following is a list of some of the potential hazards for helicopters:

- Settling with power, also known as a vortex ring state, is when the aircraft is unable to arrest its descent due to the rotor's downwash interfering with the aerodynamics of the rotor.

- Retreating blade stall is experienced during high speed flight and is the most common limiting factor of a helicopter's forward speed.

- Ground resonance is a self-reinforcing vibration that occurs when the lead/lag spacing of the blades of an articulated rotor system becomes irregular.

- Low-G condition is an abrupt change from a positive G-force state to a negative G-force state that results in loss of lift (unloaded disc) and subsequent roll over. If aft cyclic is applied while the disc is unloaded, the main rotor could strike the tail causing catastrophic failure.

- Dynamic rollover in which the helicopter pivots around one of the skids and 'pulls' itself onto its side (almost like a fixed-wing aircraft ground loop).

- Powertrain failures, especially those that occur within the shaded area of the height-velocity diagram.

- Tail rotor failures which occur from either a mechanical malfunction of the tail rotor control system or a loss of tail rotor thrust authority, called "loss of tail-rotor effectiveness" (LTE).

- Brownout in dusty conditions or whiteout in snowy conditions.

- Low rotor RPM, or "rotor droop", is when the engine cannot drive the blades at sufficient RPM to maintain flight.

- Rotor overspeed, which can over-stress the rotor hub pitch bearings (brinelling) and, if severe enough, cause blade separation from the aircraft.

- Wire and tree strikes due to low altitude operations and take-offs and landings in remote locations.

- Controlled flight into terrain in which the aircraft is flown into the ground unintentionally due to a lack of situational awareness.

- Mast bumping in some helicopters

World Records

- Backpack helicopter
- Cyclogyro
- Disk loading
- Gyrodyne
- Helicopter dynamics
- Helicopter height–velocity diagram
- Helicopter manufacturer
- Helicopter Underwater Escape Training
- Jesus nut, the top central big nut that holds the rotor on
- List of helicopter airlines
- List of rotorcraft
- Monocopter
- Transverse flow effect
- Utility helicopter
- Wire strike protection system, "WSPS" for helicopters.

Powered Aircraft

An Airbus A320 powered aircraft

A powered aircraft is an aircraft that uses onboard propulsion with mechanical power generated by an aircraft engine of some kind.

Aircraft propulsion nearly always uses either a type of propeller, or a form of jet propulsion. Other potential propulsion techniques such as ornithopters are very rarely used.

Methods of Propulsion

Propeller Aircraft

A turboprop-engined Tupolev Tu-95

A propeller or airscrew comprises a set of small, wing-like aerofoil *blades* set around a central hub which spins on an axis aligned in the direction of travel. The blades are set at a *pitch* angle to the airflow, which may be fixed or variable, such that spinning the propeller creates aerodynamic lift, or *thrust*, in a forward direction.

A *tractor* design mounts the propeller in front of the power source, while a *pusher* design mounts it behind. Although the pusher design allows cleaner airflow over the wing, tractor configuration is more common because it allows cleaner airflow to the propeller and provides a better weight distribution.

Contra-rotating propellers have one propeller close behind another on the same axis, but rotating in the opposite direction.

A variation on the propeller is to use many broad blades to create a fan. Such fans are usually surrounded by a ring-shaped fairing or duct, as *ducted fans*.

Many kinds of power plant have been used to drive propellers.

The earliest designs used man power to give dirigible balloons some degree of control, and go back to Jean-Pierre Blanchard in 1784. Attempts to achieve heavier-than-air man-powered flight did not succeed fully until Paul MacCready's Gossamer Condor in 1977.

The first powered flight of an aircraft was made in a steam-powered dirigible by Henri Giffard in 1852. Attempts to marry a practical lightweight steam engine to a practical fixed-wing airframe did not succeed until much later, by which time the internal combustion engine was already dominant.

From the first controlled powered fixed-wing aircraft flight by the Wright brothers until World War II, propellers turned by the internal combustion piston engine were virtually the only type of propulsion system in use. The piston engine is still used in the majority of smaller aircraft produced, since it is efficient at the lower altitudes and slower speeds suited to propellers.

Turbine engines need not be used as jets, but may be geared to drive a propeller in the form of a turboprop. Modern helicopters also typically use turbine engines to power the rotor. Turbines provide more power for less weight than piston engines, and are better suited to small-to-medium size aircraft or larger, slow-flying types. Some turboprop designs mount the propeller directly on an engine turbine shaft, and are called propfans.

Other less common power sources include:

- Electric motors, often linked to solar panels to create a solar-powered aircraft.

- Rubber bands, wound many times to store energy, are mostly used for flying models.

Rotorcraft

Rotorcraft have spinning blades called a *rotor* which spins in the horizontal plane to provide lift. Forward thrust is usually obtained by angling the rotor disc slightly forward so that a proportion of its lift is directed backwards; these are called helicopters. Other rotorcraft are compound helicopters and autogyros which sometimes use other means of propulsion, such as propellers and jets.

The rotor of a helicopter may, like a propeller, be powered by a variety of methods such as an internal-combustion engine or jet turbine. Tip jets, fed by gases passing along hollow rotor blades from a centrally mounted engine, have been experimented with. Attempts have even been made to mount engines directly on the rotor tips.

Jet Propulsion

Jet Aircraft

Airbreathing jet engines provide thrust by taking in air, compressing the air, injecting fuel into the hot compressed air mixture in a combustion chamber, the resulting accelerated exhaust ejects rearwards through a turbine which drives the compressor. The reaction against this acceleration provides the engine thrust.

A jet-engined Boeing 777 taking off

Jet engines can provide much higher thrust than propellers, and are naturally efficient at higher altitudes, being able to operate above 40,000 ft (12,000 m). They are also much more fuel-efficient at normal flight speeds than rockets. Consequently, nearly all high-speed and high-altitude aircraft use jet engines.

The early turbojet and modern turbofan use a spinning compressor and turbine to provide thrust. Many, mostly in military aviation, add an afterburner which injects extra fuel into the hot exhaust.

Use of a turbine is not absolutely necessary: other designs include the crude pulse jet, high-speed ramjet and the still-experimental supersonic-combustion ramjet or scramjet. These mechanically simple designs require an existing airflow to work and cannot work when stationary, so they must be launched by a catapult or rocket booster, or dropped from a mother ship.

The bypass turbofan engines of the Lockheed SR-71 were a hybrid design – the aircraft took off and landed in jet turbine configuration, and for high-speed flight the afterburner was lit and the turbine bypassed, to create a ramjet.

The motorjet was a very early design which used a piston engine in place of the combustion chamber, similar to a turbocharged piston engine except that the thrust is derived from the turbine instead of the crankshaft. It was soon superseded by the turbojet and remained a curiosity.

Rocket-powered Aircraft

Rocket propulsion offers very high thrust for light weight and has no height limit, but suffers from high fuel consumption and the need to carry oxidant as well as propellant.

Rocket-powered aircraft have been experimented with, and during the Second World War the Messerschmitt *Komet* fighter was developed and used operationally. Since then they have been restricted to specialised niches, such as the Bell X-1 which broke the sound barrier or the North American X-15 which was capable of flying at extremely high altitudes at the border with space as it was not dependent on atmospheric oxygen.

Rockets have more often been used as a supplement to the main powerplant, typically in the case of rocket-assisted take off to give more power for a heavily loaded aircraft or reduce the takeoff run. In a number of designs such as the prototype "mixed-power" Saunders-Roe SR.53 interceptor a rocket was used to provide high-speed climb and speed to reach the target while a smaller turbojet provided a slower and more economical return to base.

Ornithopter

The ornithopter obtains thrust by flapping its wings. When the wing flaps, as opposed to gliding, it continues to develop lift as before, but the lift is rotated forward to provide a thrust component.

Working devices have been created for flight research and as prototypes, but the vertical oscillation of the fuselage, which tends to accompany the wing flapping, limits their usefulness. The only practical application is a flying model hawk used to freeze prey animals into stillness so that they can be captured.

Toys in the form of a flying model bird are also popular.

Methods of Powering Lift

A fixed-wing aircraft obtains lift from airflow over the wing resulting from motion due to forward thrust. A few other types, such as the rotary-winged autogyro, obtain lift through similar methods.

Some types use a separate power system to create lift. These include the rotary-winged helicopter and craft that use lift jets (e.g. the flying bedstead).

A hot air balloon requires a power source (normally a gas burner) for lift, but is not normally considered a "powered aircraft".

Fixed-wing Aircraft

A Boeing 737 aeroplane - an example of a fixed-wing aircraft

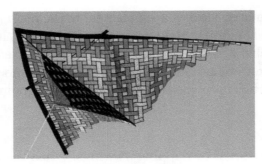
A delta-shaped kite

A fixed-wing aircraft is an aircraft, such as an aeroplane, which is capable of flight using wings that generate lift caused by the vehicle's forward airspeed and the shape of the wings. Fixed-wing aircraft are distinct from rotary-wing aircraft, in which the wings form a rotor mounted on a spinning shaft, and ornithopters, in which the wings flap in similar manner to a bird.

Glider fixed-wing aircraft, including free-flying gliders of various kinds and tethered kites, can use moving air to gain height. Powered fixed-wing aircraft that gain forward thrust from an engine (aeroplanes) include powered paragliders, powered hang gliders and some ground effect vehicles.

The wings of a fixed-wing aircraft are not necessarily rigid; kites, hang-gliders, variable-sweep wing aircraft and aeroplanes using wing-warping are all fixed-wing aircraft. Most fixed-wing aircraft are flown by a pilot on board the aircraft, but some are designed to be remotely or computer-controlled.

Early Kites

Kites were used approximately 2,800 years ago in China, where materials ideal for kite building were readily available. Some authors hold that leaf kites were being flown much earlier in what is now Indonesia, based on their interpretation of cave paintings on Muna Island off Sulawesi. By at least 549 AD paper kites were being flown, as it was recorded in that year a paper kite was used as a message for a rescue mission. Ancient and medieval Chinese sources list other uses of kites for measuring distances, testing the wind, lifting men, signaling, and communication for military operations.

Boys flying a kite in 1828 Bavaria, by Johann Michael Voltz

Stories of kites were brought to Europe by Marco Polo towards the end of the 13th century, and kites were brought back by sailors from Japan and Malaysia in the 16th and 17th centuries. Although they were initially regarded as mere curiosities, by the 18th and 19th centuries kites were being used as vehicles for scientific research.

Gliders and Powered Models

Around 400 BC in Greece, Archytas was reputed to have designed and built the first artificial, self-propelled flying device, a bird-shaped model propelled by a jet of what was probably steam, said to have flown some 200 m (660 ft). This machine may have been suspended for its flight.

Some of the earliest recorded attempts with gliders were those by the 9th-century poet Abbas Ibn Firnas and the 11th-century monk Eilmer of Malmesbury; both experiments injured their pilots.

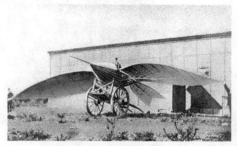

Le Bris and his glider, Albatros II, photographed by Nadar, 1868

In 1799, Sir George Cayley set forth the concept of the modern aeroplane as a fixed-wing flying machine with separate systems for lift, propulsion, and control. Cayley was building and flying models of fixed-wing aircraft as early as 1803, and he built a successful passenger-carrying glider in 1853. In 1856, Frenchman Jean-Marie Le Bris made the first powered flight, by having his glider *"L'Albatros artificiel"* pulled by a horse on a beach. In 1884, the American John J. Montgomery made controlled flights in a glider as a part of a series of gliders built between 1883-1886. Other aviators who made similar flights at that time were Otto Lilienthal, Percy Pilcher, and Octave Chanute.

In the 1890s, Lawrence Hargrave conducted research on wing structures and developed a box kite that lifted the weight of a man. His box kite designs were widely adopted. Although he also developed a type of rotary aircraft engine, he did not create and fly a powered fixed-wing aircraft.

Powered Flight

Sir Hiram Maxim built a craft that weighed 3.5 tons, with a 110-foot (34-meter) wingspan that was powered by two 360-horsepower (270-kW) steam engines driving two propellers. In 1894, his machine was tested with overhead rails to prevent it from rising. The test showed that it had enough lift to take off. The craft was uncontrollable, which Maxim, it is presumed, realized, because he subsequently abandoned work on it.

Wright Flyer III piloted by Orville Wright over Huffman Prairie, 4 October 1905

The Wright brothers' flights in 1903 with their *Flyer I* are recognized by the *Fédération Aéronautique Internationale* (FAI), the standard setting and record-keeping body for aeronautics, as "the first sustained and controlled heavier-than-air powered flight". By 1905, the Wright Flyer III was capable of fully controllable, stable flight for substantial periods.

Santos-Dumont's self-propelled 14-bis on an old postcard

In 1906, Brazilian inventor Alberto Santos Dumont designed, built and piloted an aircraft that set the first world record recognized by the Aéro-Club de France by flying the 14 bis 220 metres (720 ft) in less than 22 seconds. The flight was certified by the FAI. This was the first controlled flight, to be officially recognised, by a plane able to take off under its own power alone without any auxiliary machine such as a catapult.

The Bleriot VIII design of 1908 was an early aircraft design that had the modern mono-plane tractor configuration. It had movable tail surfaces controlling both yaw and pitch, a form of roll control supplied either by wing warping or by ailerons and controlled by its pilot with a joystick and rudder bar. It was an important predecessor of his later Bleriot XI Channel-crossing aircraft of the summer of 1909.

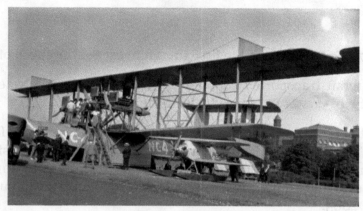

Curtiss NC-4 flying boat after it completed the first crossing of the Atlantic by a fixed-wing heavier-than-air aircraft in 1919.

World War I

World War I served as a testbed for the use of the aircraft as a weapon. Aircraft demon-strated their potential as mobile observation platforms, then proved themselves to be machines of war capable of causing casualties to the enemy. The earliest known aerial victory with a synchronised machine gun-armed fighter aircraft occurred in 1915, by German Luftstreitkräfte *Leutnant* Kurt Wintgens. Fighter aces appeared; the greatest (by number of air victories) was Manfred von Richthofen.

Following WWI, aircraft technology continued to develop. Alcock and Brown crossed the Atlantic non-stop for the first time in 1919. The first commercial flights took place between the United States and Canada in 1919.

World War II

Aeroplanes had a presence in all the major battles of World War II. They were an essen-tial component of the military strategies of the period, such as the German Blitzkrieg or the American and Japanese aircraft carrier campaigns of the Pacific.

Military gliders were developed and used in several campaigns, but they did not become widely used due to the high casualty rate often encountered. The Focke-Achgelis Fa 330 *Bachstelze* (Wagtail) rotor kite of 1942 was notable for its use by German submarines.

Before and during the war, both British and German designers were developing jet en-gines to power aeroplanes. The first jet aircraft to fly, in 1939, was the German Heinkel

He 178. In 1943 the first operational jet fighter, the Messerschmitt Me 262, went into service with the German Luftwaffe and later in the war the British Gloster Meteor entered service but never saw action — top airspeeds of aircraft for that era went as high as 1,130 km/h (702 mph), with the early July 1944 unofficial record flight of the German Me 163B V18 rocket fighter prototype.

Postwar

In October 1947, the Bell X-1 was the first aircraft to exceed the speed of sound.

In 1948–49, aircraft transported supplies during the Berlin Blockade. New aircraft types, such as the B-52, were produced during the Cold War.

The first jet airliner, the de Havilland Comet, was introduced in 1952, followed by the Soviet Tupolev Tu-104 in 1956. The Boeing 707, the first widely successful commercial jet, was in commercial service for more than 50 years, from 1958 to 2010. The Boeing 747 was the world's biggest passenger aircraft from 1970 until it was surpassed by the Airbus A380 in 2005.

Classes of Fixed-wing Aircraft

Airplane/Aeroplane

An aeroplane (also known as an airplane or simply a plane) is a powered fixed-wing aircraft that is propelled forward by thrust from a jet engine or propeller. Planes come in a variety of sizes, shapes, and wing configurations. The broad spectrum of uses for planes includes recreation, transportation of goods and people, military, and research.

Seaplane

A seaplane is a fixed-wing aircraft capable of taking off and landing (alighting) on water. Seaplanes that can also operate from dry land are a subclass called amphibian aircraft. These aircraft were sometimes called hydroplanes. Seaplanes and amphibians are usually divided into two categories based on their technological characteristics: floatplanes and flying boats.

- A floatplane is similar in overall design to a land-based aeroplane, with a generally unmodified fuselage from as compared to its landplane version, except that the wheels at the base of the undercarriage are replaced by floats, allowing the craft to operate from water rather than from dry land.

- A flying boat is a seaplane with a watertight hull forming the lower (ventral) areas of its fuselage, resting directly on the water's surface. It differs from a float plane as it does not need additional floats for buoyancy, although it may have small underwing floats or fuselage-mount sponsons to stabilize it on the water.

Large seaplanes are usually flying boats, with most classic amphibian aircraft designs using some form of flying-boat design for their fuselage/hull.

Powered Gliders

Many forms of glider may be modified by adding a small power plant. These include:

- Motor glider - a conventional glider or sailplane with an auxiliary power plant that may be used when in flight to increase performance.

- Powered hang glider - a hang glider with a power plant added.

- Powered parachute - a paraglider type of parachute with an integrated airframe, seat, undercarriage and power plant hung beneath.

- Powered paraglider or paramotor - a paraglider with a power plant suspended behind the pilot.

Ground Effect Vehicle

A ground effect vehicle (GEV) is a craft that attains level flight near the surface of the earth, making use of the ground effect – an aerodynamic interaction between the wings and the earth's surface. Some GEVs are able to fly higher out of ground effect (OGE) when required – these are classed as powered fixed-wing aircraft.

Glider

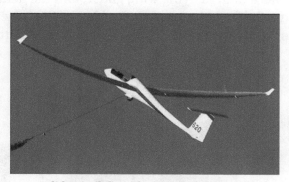

A glider (sailplane) being winch-launched

A glider is a heavier-than-air craft that is supported in flight by the dynamic reaction of the air against its lifting surfaces, and whose free flight does not depend on an engine. A sailplane is a fixed-wing glider designed for soaring - the ability to gain height in updrafts of air and to fly for long periods.

Gliders are mainly used for recreation, but have also been used for other purposes such as aerodynamics research, warfare and recovering spacecraft.

A motor glider does have an engine for extending its performance and some have engines powerful enough to take off, but the engine is not used in normal flight.

As is the case with planes, there are a wide variety of glider types differing in the construction of their wings, aerodynamic efficiency, location of the pilot and controls. Perhaps the most familiar type is the toy paper plane.

Large gliders are most commonly launched by a tow-plane or by a winch. Military gliders have been used in war to deliver assault troops, and specialised gliders have been used in atmospheric and aerodynamic research. Rocket-powered aircraft and space-planes have also made unpowered landings.

Gliders and sailplanes that are used for the sport of gliding have high aerodynamic efficiency. The highest lift-to-drag ratio is 70:1, though 50:1 is more common. After launch, further energy is obtained through the skillful exploitation of rising air in the atmosphere. Flights of thousands of kilometres at average speeds over 200 km/h have been achieved.

The most numerous unpowered aircraft are paper aeroplanes, a handmade type of glider. Like hang gliders and paragliders, they are foot-launched and are in general slower, smaller, and less expensive than sailplanes. Hang gliders most often have flexible wings given shape by a frame, though some have rigid wings. Paragliders and paper aeroplanes have no frames in their wings.

Gliders and sailplanes can share a number of features in common with powered aircraft, including many of the same types of fuselage and wing structures. For example, the Horten H.IV was a tailless flying wing glider, and the delta wing-shaped Space Shuttle orbiter flew much like a conventional glider in the lower atmosphere. Many gliders also use similar controls and instruments as powered craft.

Types of Glider

(video) A glider sails over Gunma, Japan.

The main application today of glider aircraft is sport and recreation.

Sailplane

Gliders were developed from the 1920s for recreational purposes. As pilots began to understand how to use rising air, sailplane gliders were developed with a high lift-to-drag ratio. These allowed longer glides to the next source of 'lift', and so increase their chances of flying long distances. This gave rise to the popular sport of gliding.

Early gliders were mainly built of wood and metal but the majority of sailplanes now use composite materials incorporating glass, carbon or aramid fibres. To minimise drag, these types have a streamlined fuselage and long narrow wings having a high aspect ratio. Both single-seat and two-seat gliders are available.

Initially training was done by short 'hops' in primary gliders which are very basic aircraft with no cockpit and minimal instruments. Since shortly after World War II training has always been done in two-seat dual control gliders, but high performance two-seaters are also used to share the workload and the enjoyment of long flights. Originally skids were used for landing, but the majority now land on wheels, often retractable. Some gliders, known as motor gliders, are designed for unpowered flight, but can deploy piston, rotary, jet or electric engines. Gliders are classified by the FAI for competitions into glider competition classes mainly on the basis of span and flaps.

Ultralight "airchair" Goat 1 glider

A class of ultralight sailplanes, including some known as microlift gliders and some as 'airchairs', has been defined by the FAI based on a maximum weight. They are light enough to be transported easily, and can be flown without licensing in some countries. Ultralight gliders have performance similar to hang gliders, but offer some additional crash safety as the pilot can be strapped in an upright seat within a deformable structure. Landing is usually on one or two wheels which distinguishes these craft from hang gliders. Several commercial ultralight gliders have come and gone, but most current development is done by individual designers and home builders.

Military Gliders

Military gliders were used during World War II for carrying troops (glider infantry) and heavy equipment to combat zones. The gliders were towed into the air and most of the way to their target by military transport planes, e.g. C-47 Dakota, or by bombers that had been relegated to secondary activities, e.g. Short Stirling. Once released from the tow near the target, they landed as close to the target as possible. The advantage over paratroopers were that heavy equipment could be landed and that the troops were quickly assembled rather than being dispersed over a drop zone. The gliders were treated as disposable, leading to construction from common and inexpensive materials such as wood, though a few were retrieved and re-used. By the time of the Korean War, transport aircraft had also become larger and more efficient so that even light tanks could be dropped by parachute, causing gliders to fall out of favor.

A Waco CG-4A of the USAAF in 1943

Research Gliders

Even after the development of powered aircraft, gliders continued to be used for aviation research. The NASA Paresev Rogallo flexible wing was originally developed to investigate alternative methods of recovering spacecraft. Although this application was abandoned, publicity inspired hobbyists to adapt the flexible-wing airfoil for modern hang gliders.

Initial research into many types of fixed-wing craft, including flying wings and lifting bodies was also carried out using unpowered prototypes.

Hang Glider

Hang gliding

A hang glider is a glider aircraft in which the pilot is ensconced in a harness suspended from the airframe, and exercises control by shifting body weight in opposition to a control frame. Most modern hang gliders are made of an aluminium alloy or composite-framed fabric wing. Pilots have the ability to soar for hours, gain thousands of metres of altitude in thermal updrafts, perform aerobatics, and glide cross-country for hundreds of kilometres.

Paraglider

A paraglider is a lightweight, free-flying, foot-launched glider aircraft with no rigid primary structure. The pilot sits in a harness suspended below a hollow fabric wing whose shape is formed by its suspension lines, the pressure of air entering vents in the front of the wing and the aerodynamic forces of the air flowing over the outside. Paragliding is most often a recreational activity.

Unmanned Gliders

A paper plane is a toy aircraft (usually a glider) made out of paper or paperboard.

Model glider aircraft are models of aircraft using lightweight materials such as polystyrene and balsa wood. Designs range from simple glider aircraft to accurate scale models, some of which can be very large.

Glide bombs are bombs with aerodynamic surfaces to allow a gliding flightpath rather than a ballistic one. This enables the carrying aircraft to attack a heavily defended target from a distance.

Kite

A kite in flight

A kite is an aircraft tethered to a fixed point so that the wind blows over its wings. Lift is generated when air flows over the kite's wing, producing low pressure above the wing and high pressure below it, and deflecting the airflow downwards. This deflection also generates horizontal drag in the direction of the wind. The resultant force vector from the lift and drag force components is opposed by the tension of the one or more rope lines or tethers attached to the wing.

Kites are mostly flown for recreational purposes, but have many other uses. Early pioneers such as the Wright Brothers and J.W. Dunne sometimes flew an aircraft as a kite in order to develop it and confirm its flight characteristics, before adding an engine and flight controls, and flying it as an aeroplane.

Uses

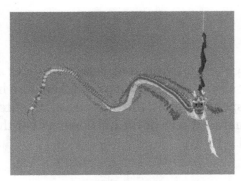

Chinese dragon kite more than one hundred feet long which flew in the Berkeley, California, kite festival in 2000

Military Applications

Kites have been used for signaling, for delivery of munitions, and for observation, by lifting an observer above the field of battle, and by using kite aerial photography.

Science and Meteorology

Kites have been used for scientific purposes, such as Benjamin Franklin's famous experiment proving that lightning is electricity. Kites were the precursors to the traditional aircraft, and were instrumental in the development of early flying craft. Alexander Graham Bell experimented with very large man-lifting kites, as did the Wright brothers and Lawrence Hargrave. Kites had a historical role in lifting scientific instruments to measure atmospheric conditions for weather forecasting.

Radio Aerials and Light Beacons

Kites can be used to carry radio antennas. This method was used for the reception station of the first transatlantic transmission by Marconi. Captive balloons may be more convenient for such experiments, because kite-carried antennas require a lot of wind, which may be not always possible with heavy equipment and a ground conductor.

Kites can be used to carry light effects such as lightsticks or battery powered lights.

Kite Traction

Kites can be used to pull people and vehicles downwind. Efficient foil-type kites such as power kites can also be used to sail upwind under the same principles as used by other sailing craft, provided that lateral forces on the ground or in the water are redirected as with the keels, center boards, wheels and ice blades of traditional sailing craft. In the last two decades several kite sailing sports have become popular, such as kite buggying, kite landboarding, kite boating and kite surfing. Snow kiting has also become popular.

A quad-line traction kite, commonly used as a power source for kite surfing

Kite sailing opens several possibilities not available in traditional sailing:

- Wind speeds are greater at higher altitudes

- Kites may be manoeuvered dynamically which increases the force available dramatically

- There is no need for mechanical structures to withstand bending forces; vehicles or hulls can be very light or dispensed with all together

Power Generation

Conceptual research and development projects by over a hundred entities are investigating the use of kites in harnessing high altitude wind currents to generate electricity.

Cultural Uses

Kite festivals are a popular form of entertainment throughout the world. They include local events, traditional festivals and major international festivals.

Designs

Train of connected kites

- Bermuda kite

- Bowed kite, e.g. Rokkaku

- Cellular or box kite

- Chapi-chapi

- Delta kite

- Foil, parafoil or bow kite

- Malay kite

- Tetrahedral kite

Types

- Expanded polystyrene kite

- Fighter kite

- Indoor kite

- Inflatable single-line kite

- Kytoon

- Man-lifting kite

- Rogallo parawing kite

- Stunt (sport) kite

- Water kite

Characteristics

An IAI Heron - an unmanned aerial vehicle with a twin-boom configuration

Airframe

The structural parts of a fixed-wing aircraft are called the airframe. The parts present can vary according to the aircraft's type and purpose. Early types were usually made of wood with fabric wing surfaces, When engines became available for powered flight around a hundred years ago, their mounts were made of metal. Then as speeds increased more and more parts became metal until by the end of WWII all-metal aircraft were common. In modern times, increasing use of composite materials has been made.

Typical structural parts include:

- One or more large horizontal *wings*, often with an airfoil cross-section shape. The wing deflects air downward as the aircraft moves forward, generating lifting force to support it in flight. The wing also provides stability in roll to stop the aircraft from rolling to the left or right in steady flight.

The An-225 Mriya, which can carry a 250-tonne payload, has two vertical stabilisers.

- A *fuselage*, a long, thin body, usually with tapered or rounded ends to make its shape aerodynamically smooth. The fuselage joins the other parts of the airframe and usually contains important things such as the pilot, payload and flight systems.

- A *vertical stabiliser* or fin is a vertical wing-like surface mounted at the rear of the plane and typically protruding above it. The fin stabilises the plane's yaw (turn left or right) and mounts the rudder which controls its rotation along that axis.

- A *horizontal stabiliser*, usually mounted at the tail near the vertical stabilizer. The horizontal stabilizer is used to stabilise the plane's pitch (tilt up or down) and mounts the elevators which provide pitch control.

- *Landing gear,* a set of wheels, skids, or floats that support the plane while it is on the surface. On seaplanes the bottom of the fuselage or floats (pontoons) support it while on the water. On some planes the landing gear retracts during flight to reduce drag.

Wings

The wings of a fixed-wing aircraft are static planes extending either side of the aircraft. When the aircraft travels forwards, air flows over the wings which are shaped to create lift.

Wing Structure

Kites and some light weight gliders and aeroplanes have flexible wing surfaces which are stretched across a frame and made rigid by the lift forces exerted by the airflow over them. Larger aircraft have rigid wing surfaces which provide additional strength.

Whether flexible or rigid, most wings have a strong frame to give them their shape and to transfer lift from the wing surface to the rest of the aircraft. The main structural elements are one or more spars running from root to tip, and many ribs running from the leading (front) to the trailing (rear) edge.

Early aeroplane engines had little power and light weight was very important. Also, early aerofoil sections were very thin, and could not have strong frame installed within. So until the 1930s most wings were too light weight to have enough strength and external bracing struts and wires were added. When the available engine power increased during the 1920s and 1930s, wings could be made heavy and strong enough that bracing was not needed any more. This type of unbraced wing is called a cantilever wing.

Wing Configuration

Captured Morane-Saulnier L wire-braced parasol monoplane

The number and shape of the wings varies widely on different types. A given wing plane may be full-span or divided by a central fuselage into port (left) and starboard (right) wings. Occasionally even more wings have been used, with the three-winged triplane achieving some fame in WWI. The four-winged quadruplane and other Multiplane (aeronautics) designs have had little success.

A monoplane has a single wing plane, a biplane has two stacked one above the other, a tandem wing has two placed one behind the other. When the available engine power increased during the 1920s and 1930s and bracing was no longer needed, the unbraced or cantilever monoplane became the most common form of powered type.

The wing planform is the shape when seen from above. To be aerodynamically efficient, a wing should be straight with a long span from side to side but have a short chord (high aspect ratio). But to be structurally efficient, and hence light weight, a wing must have a short span but still enough area to provide lift (low aspect ratio).

At transonic speeds, near the speed of sound, it helps to sweep the wing backwards or forwards to reduce drag from supersonic shock waves as they begin to form. The swept wing is just a straight wing swept backwards or forwards.

Two Dassault Mirage G prototypes, one with wings swept

The delta wing is a triangle shape which may be used for a number of reasons. As a flexible Rogallo wing it allows a stable shape under aerodynamic forces, and so is often used for kites and other ultralight craft. As a supersonic wing it combines high strength with low drag and so is often used for fast jets.

A variable geometry wing can be changed in flight to a different shape. The variable-sweep wing transforms between an efficient straight configuration for takeoff and landing, to a low-drag swept configuration for high-speed flight. Other forms of variable planform have been flown, but none have gone beyond the research stage.

Fuselage

A *fuselage* is a long, thin body, usually with tapered or rounded ends to make its shape aerodynamically smooth. The fuselage may contain the flight crew, passengers, cargo or payload, fuel and engines. The pilots of manned aircraft operate them from a *cockpit* located at the front or top of the fuselage and equipped with controls and usually windows and instruments. A plane may have more than one fuselage, or it may be fitted with booms with the tail located between the booms to allow the extreme rear of the fuselage to be useful for a variety of purposes.

Wings vs. Bodies

Flying Wing

The US-produced B-2 Spirit, a strategic bomber using a flying wing configuration which is capable of intercontinental missions

A flying wing is a tailless aircraft which has no definite fuselage, with most of the crew, payload and equipment being housed inside the main wing structure.

The flying wing configuration was studied extensively in the 1930s and 1940s, notably by Jack Northrop and Cheston L. Eshelman in the United States, and Alexander Lippisch and the Horten brothers in Germany. After the war, a number of experimental designs were based on the flying wing concept. Some general interest continued until the early 1950s but designs did not necessarily offer a great advantage in range and presented a number of technical problems, leading to the adoption of "conventional" solutions like the Convair B-36 and the B-52 Stratofortress. Due to the practical need for a deep wing, the flying wing concept is most practical for designs in the slow-to-medium speed range, and there has been continual interest in using it as a tactical airlifter design.

Interest in flying wings was renewed in the 1980s due to their potentially low radar reflection cross-sections. Stealth technology relies on shapes which only reflect radar waves in certain directions, thus making the aircraft hard to detect unless the radar receiver is at a specific position relative to the aircraft - a position that changes continuously as the aircraft moves. This approach eventually led to the Northrop B-2 Spirit stealth bomber. In this case the aerodynamic advantages of the flying wing are not the primary needs. However, modern computer-controlled fly-by-wire systems allowed for many of the aerodynamic drawbacks of the flying wing to be minimised, making for an efficient and stable long-range bomber.

Blended Wing Body

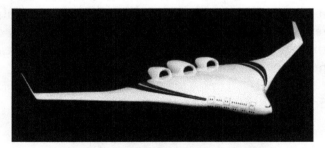

Computer-generated model of the Boeing X-48

Blended wing body aircraft have a flattened and airfoil shaped body, which produces most of the lift to keep itself aloft, and distinct and separate wing structures, though the wings are smoothly blended in with the body.

Thus blended wing bodied aircraft incorporate design features from both a futuristic fuselage and flying wing design. The purported advantages of the blended wing body approach are efficient high-lift wings and a wide airfoil-shaped body. This enables the entire craft to contribute to lift generation with the result of potentially increased fuel economy.

Lifting Body

The Martin Aircraft Company X-24 was built as part of a 1963 to 1975 experimental US military program.

A lifting body is a configuration in which the body itself produces lift. In contrast to a flying wing, which is a wing with minimal or no conventional fuselage, a lifting body can be thought of as a fuselage with little or no conventional wing. Whereas a flying wing seeks to maximize cruise efficiency at subsonic speeds by eliminating non-lifting surfaces, lifting bodies generally minimize the drag and structure of a wing for subsonic, supersonic, and hypersonic flight, or, spacecraft re-entry. All of these flight regimes pose challenges for proper flight stability.

Lifting bodies were a major area of research in the 1960s and 1970s as a means to build a small and lightweight manned spacecraft. The US built a number of famous lifting body rocket planes to test the concept, as well as several rocket-launched re-entry vehicles that were tested over the Pacific. Interest waned as the US Air Force lost interest in the manned mission, and major development ended during the Space Shuttle design process when it became clear that the highly shaped fuselages made it difficult to fit fuel tankage.

Empennage and Foreplane

The classic aerofoil section wing is unstable in flight and difficult to control. Flexible-wing types often rely on an anchor line or the weight of a pilot hanging beneath to maintain the correct attitude. Some free-flying types use an adapted aerofoil that is stable, or other ingenious mechanisms including, most recently, electronic artificial stability.

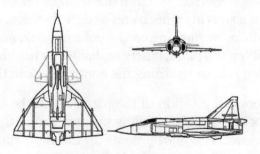

Canards on the Saab Viggen

But in order to achieve trim, stability and control, most fixed-wing types have an empennage comprising a fin and rudder which act horizontally and a tailplane and elevator which act vertically. This is so common that it is known as the conventional layout. Sometimes there may be two or more fins, spaced out along the tailplane.

Some types have a horizontal "canard" foreplane ahead of the main wing, instead of behind it. This foreplane may contribute to the trim, stability or control of the aircraft, or to several of these.

Aircraft Controls

Kite Control

Kites are controlled by wires running down to the ground. Typically each wire acts as a tether to the part of the kite it is attached to.

Free-flying Aircraft Controls

Gliders and aeroplanes have more complex control systems, especially if they are piloted.

Typical light aircraft (Cessna 150M) cockpit with control yokes

The main controls allow the pilot to direct the aircraft in the air. Typically these are:

- The *yoke* or *joystick* controls rotation of the plane about the pitch and roll axes. A yoke resembles a steering wheel, and a control stick is a joystick. The pilot can pitch the plane down by pushing on the yoke or stick, and pitch the plane up by pulling on it. Rolling the plane is accomplished by turning the yoke in the direction of the desired roll, or by tilting the control stick in that direction.

- *Rudder pedals* control rotation of the plane about the yaw axis. There are two pedals that pivot so that when one is pressed forward the other moves backward, and vice versa. The pilot presses on the right rudder pedal to make the plane yaw to the right, and pushes on the left pedal to make it yaw to the left.

The rudder is used mainly to balance the plane in turns, or to compensate for winds or other effects that tend to turn the plane about the yaw axis.

- On powered types, an engine stop control ("fuel cutoff", for example) and, usually, a *Throttle* or *thrust lever* and other controls, such as a fuel-mixture control (to compensate for air density changes with altitude change).

Other common controls include:

- *Flap levers,* which are used to control the deflection position of flaps on the wings.

- *Spoiler levers,* which are used to control the position of spoilers on the wings, and to arm their automatic deployment in planes designed to deploy them upon landing. The spoilers reduce lift for landing.

- *Trim controls,* which usually take the form of knobs or wheels and are used to adjust pitch, roll, or yaw trim. These are often connected to small airfoils on the trail edge of the control surfaces called 'trim tabs'. Trim is used to reduce the amount of pressure on the control forces needed to maintain a steady course.

- On wheeled types, *Brakes* are used to slow and stop the plane on the ground, and sometimes for turns on the ground.

A craft may have two pilots' seats with dual controls, allowing two pilots to take turns. This is often used for training or for longer flights.

The control system may allow full or partial automation of flight, such as an autopilot, a wing leveler, or a flight management system. An unmanned aircraft has no pilot but is controlled remotely or via means such as gyroscopes or other forms of autonomous control.

Cockpit Instrumentation

Six basic flight instruments

On manned types, instruments provide information to the pilots, including flight, engines, navigation, communications and other aircraft systems that may be installed.

The six basic instruments (sometimes referred to as the six pack) include:

- An *airspeed indicator,* which indicates the speed at which the plane is moving through the surrounding air.

- An *altimeter,* which indicates the altitude or height of the plane above mean sea level.

- A *heading indicator,* (sometimes referred to as a "directional gyro (DG)"), which indicates the magnetic compass heading that the plane's fuselage is pointing towards. The actual direction the plane is flying towards is affected by the wind conditions.

- An *attitude indicator,* sometimes called an *artificial horizon,* which indicates the exact orientation of the plane about its pitch and roll axes.

- A *vertical speed indicator,* which shows the rate at which the plane is climbing or descending.

- A *turn coordinator,* or *turn and bank indicator* which helps the pilot maintain the plane in a coordinated attitude while turning.

Other instruments might include:

- A *two-way radio* to enable communications with other planes and air traffic control. Planes built before World War II may not have been equipped with a radio but they are nearly essential now.

- A *horizontal situation indicator,* shows the position and movement of the plane as seen from above with respect to the ground, including course/heading and other information.

- Instruments showing the status of each engine in the plane (operating speed, thrust, temperature, RPM, and other variables).

- Combined display systems such as *primary flight displays* or *navigation displays.*

- Information displays such as on-board *weather radar* displays.

- A *radio direction finder* which indicates the direction to one or more radio beacons and which can be used to determine the plane's position.

- A *satellite navigation* system to provide an accurate position.

References

- Crouch, Tom (1982). Bleriot XI, The Story of a Classic Aircraft. Smithsonian Institution Press. pp. 21 and 22. ISBN 0-87474-345-1.

- Schweizer, Paul A: Wings Like Eagles, The Story of Soaring in the United States, pages 14-22. Smithsonian Institution Press, 1988. ISBN 0-87474-828-3

- Crane, Dale: Dictionary of Aeronautical Terms, third edition. Aviation Supplies & Academics, 1997. ISBN 1-56027-287-2

- de Bie, Rob. "Me 163B Komet - Me 163 Production - Me 163B: Werknummern list." robdebie. home. Retrieved: 28 July 2013.

Permissions

Index

www.ingramcontent.com/pod-product-compliance
Lightning Source LLC
Jackson TN
JSHW052154130125
77033JS00004B/179